CONTENTS

ISBN: 9780170368186

 ISBN: 9780170368186

Glossary

Make your own glossary of key terms:

Term	Definition	Picture/Example
BEDMAS		
Factor		
Multiple		
Highest common factor (HCF)		
Lowest common multiple (LCM)		
Prime number		
Integer		
Root		
Negative power		
Decimal place		

ISBN: 9780170368186

Term	Definition	Picture/Example
Significant figure		
Fraction		
Reciprocal		
Percentage		
Interest		
Ratio		
Proportion		
Exchange rate		
Standard form (scientific notation)		
Simple interest (SI)		
Compound interest (CI)		

ISBN: 9780170368186

Order of operations

When simplifying and solving expressions, we use **BEDMAS** to remind us of the order of operations.

Brackets	$(2 + 3) - 1$
Exponents	3^2
Division	$10 \div 5$
Multiplication	2×6
Addition	$2 + 3$
Subtraction	$4 - 1$

Example:

$20 - 2(5 - 2)^2 + 1$	**Brackets**	$(5 - 2) = 3$
$= 20 - 2 \times 3^2 + 1$	**Exponent**	$3^2 = 9$
$= 20 - 2 \times 9 + 1$	**Multiplication**	$2 \times 9 = 18$
$= 20 - 18 + 1$	**Addition**	$-18 + 1 = -17$
$= 20 - 17$	**Subtraction**	$20 - 17 = 3$
$= 3$		

Calculate the values of the following.

1 $6 - 2 \times 2$

2 $2 \times 8 - \frac{12}{2}$

3 $2(4 + 1)$

4 $8 - 2(4 - 1)$

5 2×4^2

6 $\frac{3^2}{2^3}$

7 $(6 \times 9) + (6 + 9)$

8 $2^2 + (6 - 1)$

9 $\frac{3 + 4}{2 \times 2}$

10 $\frac{3^2 + 1}{4 \div 2}$

ISBN: 9780170368186

11 $(6 + 2)^2$

12 $(2 \times 3 - 1)^3$

13 $\frac{8 + (5 - 2)^2 - 1}{2}$

14 $3 + 3 \div 3 \times 3 - 3$

15 $3 + 2 \times 4^2 \div 8 - 5$

16 $\frac{3(6 - 2^2)^2 + 8}{5}$

17 $\frac{20 - 3 \times 2^2 + (5 - 1)^2}{6}$

18 $\frac{2^3 \times 7\ (7 - 4) + 8}{2(3 - 1)^2}$

19 Olivia has 14 lollies. She eats half and then gives two away. How many does she have left?

20 Hemi has 36 lollies. He shares these equally between himself and two friends. Then he eats five. How many does he have left?

21 Aroha has 16 lollies and Amelia has 11. They combine their lollies, and then share them equally between themselves and Anna. If Anna then eats two of hers, how many does she have left?

22 Max earned $20 for mowing the neighbour's lawn. He bought movie tickets at $7 each for himself and his sister, an ice cream for $4, and then found a $1 coin in his jacket pocket. How much money does he have?

ISBN: 9780170368186

Using your calculator

Do the calculations on the next page using your calculator, then turn your calculator upside down and add the word to the story below. Be careful to check that it makes sense.

Example: $103^2 \times 5 = 53045$; upside down this spells SHOES.

Once upon a time ...

Once upon a time there was a 1 ________________ named 2 ________________.

3 ________________ princes would 4 ________________ her castle, which was located on a 5 ________________ in the middle of a swamp. They would 6 ________________ and 7 ________________ through 8 ________________ filled with 9 ________________, until the 10 ________________ of their 11 ________________ were caked with 12 ________________ of 13 ________________ 14 ________________.

Ellie kept pet 15 ________________, which would 16 ________________ their food, then 17 ________________ and 18 ________________ about the palace. They would 19 ________________ at the princes.

However, the sight of Ellie would make the princes 20 ________________, and dream of eternal 21 ________________.

They would 22 ________________ for her hand. But she would just 23 ________________ and reject them. They would 24 ________________ and 25 ________________, and turn on their 26 ________________ and depart.

Prince 27 ________________ was different. He brought a bunch of 28 ________________ and an 29 ________________ 30 ________________. '31 ________________,' she said. '32 ________________, none of the other 33 ________________ brought me an 34 ________________ 35 ________________. I can 36 ________________ that you're no 37 ________________. You can stay and 38 ________________ with me.'

And they lived happily ever after.

ISBN: 9780170368186

1 $2(136^2 + 373)$

2 $21(12^3 - 215)$

3 $3^3 \times (1.5 \times 10^6 - 99401)$

4 $\frac{7}{6}(42^4) + \frac{5}{4}(10^3) - 5^2 + 1$

5 $(1 + 3^2(1 + 3^2(1 + 3^2(1 + 3^2)))) + 333$

6 $\frac{300(5^4 - \sqrt{576})}{4}$

7 $3^5 \times 5^2$

8 $701 \times \sqrt[3]{512}$

9 $21^2 \times \frac{1.2^2 + 0.185}{1.005 - 0.88}$

10 $\frac{10^4 + 3 \times 247}{\sqrt{0.04}}$

11 $5(\sqrt{6561} + 22)^2$

12 $5.8 \times 10^4 + \frac{5.32 \times 10^2}{7}$

13 $13^6 - \frac{317\sqrt{1521}}{0.04}$

14 $\frac{100^2 - 20^2}{3}$

15 $\frac{7(7 + 13 \times 48)}{0.5^3}$

16 $\frac{2 \times 10^5 - 2 \times 10^4 + 9 \times 10^3 + 4 \times 10^2 + 3}{0.5}$

17 $\frac{3 \times 7 \times 367}{730 - 9^3}$

18 $\frac{126.8 \times 0.125}{0.3^2 - 0.04}$

19 $\frac{6 \times 919 \times (7.412 + 15.588)}{12^2 - 11^2}$

20 $6\left(\frac{2^{12}}{2^8}\right) \times (4 \times 10^3 - 77)$

21 $5 \times 10^4 + 5 \times 10^3 + 1 \times 10^2 + 7 \times 10^1 + 8 \times 10^0$

22 $2(1 + 2(1 + 2(1 + 2(1 + 2(18))))) + 32$

23 $73^3 - \frac{1612 \times 10^2}{13} - 1$

24 $\frac{6 \times 10^4 - 5}{\sqrt{169}}$

25 $\frac{10^4 - 1145}{6^2 - 5^2}$

26 $2((\sqrt[3]{4913} + 9)^3 + 9^4 + 16^3 + 434)$

27 $3 \times 10^5 + 1 \times 10^4 + 7 \times 10^3 + 7 \times 10^2 + 1 \times 10^1 + 8 \times 10^0$

28 $6 + 3(3 + 6(1 + 2 \times 3^3 \times 547))$

29 $3(2^6 \times 3^3 - 1)$

30 $(10^2 + 10 + 1)(2^3 - 2) - 3$

31 $1000^0 - 0.45^2 - \frac{15^2 + 4^2}{10000}$

32 $751(2^3 - 2)$

33 $\frac{10^6 + 3(3 \times 10^4 - 8819)}{\sqrt{0.0291 + 0.0109}}$

34 $3(10^3 + (10^3 - 3 \times 7 \times 13))$

35 $3(10^2 + 11^2)$

36 $8^2 + 7^2 \times 6 - 23$

37 $1 - \frac{9^3 + 9 \times 7}{1001 - 23^0}$

38 $2^3 \times 3^3 \times \sqrt[3]{343} \times \sqrt{121} \times \sqrt{361}$

ISBN: 9780170368186

Factors, multiples and primes

Factors

The factors of a number are all the numbers that divide exactly into it.
Example: The factors of **9** are: 1, **3**, 9.
The factors of **24** are: 1, 2, **3**, 4, 6, 8, 12, 24.

Highest common factor (HCF)

The highest common factor of two numbers is the biggest factor of both numbers.
Example: The HCF of **9** and **24** is **3** because it is the biggest factor that is on both lists.

List all the factors of the following numbers.

1 4: ______________________

2 12: ______________________

3 15: ______________________

4 36: ______________________

5 7: ______________________

6 1: ______________________

List the factors of both numbers, and write down the HCF.

7 8: ______________________

12: ______________________

HCF: ______________________

8 5: ______________________

20: ______________________

HCF: ______________________

9 18: ______________________

60: ______________________

HCF: ______________________

10 11: ______________________

21: ______________________

HCF: ______________________

11 16: ______________________

30: ______________________

HCF: ______________________

12 17: ______________________

3: ______________________

HCF: ______________________

13 Henry has 24 mint lollies and 36 fruit lollies. He wants to use all of these to make as many identical small bags of lollies as possible. How many bags can he make, and what does each contain?

____

____

14 He buys an additional 30 caramel lollies. How many identical bags can he make now, and what does each contain?

____

ISBN: 9780170368186

Multiples

The multiples of a number are the results of multiplying the number by another number.
Example: The multiples of **3** are: 3, 6, 9, **12**, 15, 18, …
The multiples of **4** are: 4, 8, **12**, 16, 20, 24, …

These are the answers to the times tables.

Lowest common multiple (LCM)

The lowest common multiple of two numbers is the smallest multiple of both numbers.
Example: The LCM of **3** and **4** is **12**, because it is the smallest multiple that is on both lists.

List the first six multiples of the following numbers.

1 2: ______ **2** 5: ______

3 9: ______ **4** 12: ______

5 10: ______ **6** 1: ______

List at least the first six multiples of both numbers, and write down the LCM.

7 2: ______
3: ______
LCM: ______

8 5: ______
3: ______
LCM: ______

9 6: ______
9: ______
LCM: ______

10 8: ______
12: ______
LCM: ______

11 10: ______
12: ______
LCM: ______

12 7: ______
9: ______
LCM: ______

13 Melino is threading green and white beads to make a pattern. She decides that every tenth bead will be green and every eighth bead will be a big one. Which bead will be the first big green one?

______

______

14 Her green and white beads, as well as being big or small, can also be round and cubic shapes. If she does the patterns in question **13** but she makes every third bead a cubic one, which bead will be the first big, green, cubic bead?

______

______

ISBN: 9780170368186

Primes

A prime number has **exactly** two factors: 1 and itself.

Example: 2 is prime because its only factors are 1 and 2
17 is prime because its only factors are 1 and 17
41 is prime because its only factors are 1 and 41
9 is not a prime because it has three factors: 1, 3 and 9

Note: 1 is **not** prime because its only factor is 1, so it does not have two factors.

Testing for 'prime-ness'

- You need to try dividing the number by every possible factor.
- This is easy for small numbers, but **very** time consuming for big numbers.

List the prime numbers between:

1 1 and 10

2 10 and 20

3 20 and 40

4 40 and 60

Mixing it up

1 List the prime factors of 12.

2 List the prime factors of 23.

3 List the prime factors of 45.

4 List the prime factors of 60.

5 A group of friends caught a shuttle for a trip to the zoo. The shuttle fare and the zoo tickets were whole numbers of dollars. It cost them a total of $143 for the shuttle and $187 for the tickets to the zoo. How many friends were there?

6 The first of the month was a Monday, and Izzie cleaned out her pet rat's cage. If she has to clean it every third day, on what date will she next clean it on a Monday?

7 The club has fewer than 30 members. The members were asked for a $1 donation when they paid their subscription. All members except one gave the $1 donation. They collected a total of $942 in subscriptions and donations. How many club members are there and how much was the subscription?

ISBN: 9780170368186

Integers

- An integer is any positive or negative whole number, or 0.
- These are best understood on a number line.

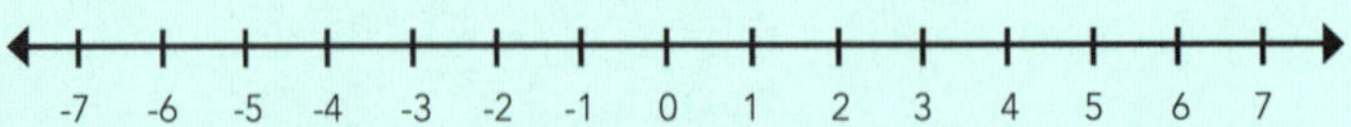

Adding and subtracting integers

- Do this on the number line.
- Locate the first number on the number line.
- Then move left or right as appropriate.

Example: -3 + 5 – 9 = -7

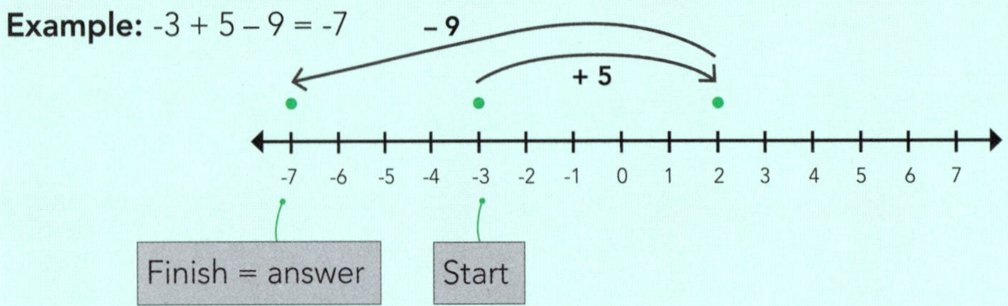

Multiplying and dividing integers

x/÷	+	–
+	+	–
–	–	+

Same signs ⇒ +
Different signs ⇒ –

Example: -3 x 8 ÷ -2 = -3 x (8 ÷ -2)
= -3 x -4
= 12

Calculations with integers.

1 3 x 4 x -1

____

2 -3 + 5

____

3 -2 – -2

____

4 4 + -3 x -2

____

5 $(-2)^2$

____

6 $\frac{-14}{-2}$

____

7 $\frac{-3 \times -4}{-2}$

____

8 $\frac{-2 \times 6 \times 1}{-2 \times -2}$

____

ISBN: 9780170368186

9 $(-3)^2$

10 $-4 \times -2 + 3 \times -1$

11 $-(-5)^2$

12 $\sqrt{25}$

13 $\frac{-2 \times -3}{-2 + 1}$

14 $\frac{-(2 - 3)^2}{5 - (3 - 1)^2}$

15 Cade had $32 in his bank account. If he spent $45, what is the new balance of his account?

16 When Lucy got up in the morning it was -2°C. By lunchtime the temperature had increased by 15°C. What was the temperature at lunchtime?

17 Ice creams are stored at a temperature of -18°C. Howard accidentally turned the freezer off and five hours later the freezer temperature has increased by 10°C. What is the new temperature of the freezer?

18 In Jeremy's building, the laundry is in the basement (i.e. floor -1). If he needs to go up five floors to visit a neighbour and then down two to get back to his apartment, what floor does he live on?

19 On a frosty morning in Twizel, the temperature at dawn was -5 °C. The temperature had increased to 12° by lunchtime. By how many degrees did the temperature increase?

20 The top of Mt McKinley, the highest point in the USA, is 6168 m above sea level. The lowest point in the USA is in Death Valley, at 86 m below sea level. Calculate the height difference between these.

21 A diver spends time at 21 m below the surface, then he ascends 13 m. How far below the surface is he now?

22 The city of Rome was founded in 753 BC. In AD 64, Nero set fire to it. How old was the city at this time?

23a Cleopatra was born in the year 69 BC and died when she was 39 years old. In what year did she die?

b Cleopatra became Queen of Egypt in 51 BC. How old was she when she became queen?

ISBN: 9780170368186

Powers and roots

Powers

Powers are used to indicate how many times a number is multiplied by itself.

The **power** indicates how many times you need to multiply.

Make sure you use **brackets** on your calculator.

Example: $3^2 = 3 \times 3$
$= 9$

$(-4)^3 = -4 \times -4 \times -4$
$= -64$

You can do this on your calculator with either x^2 or x^3 or y^x or ^ or $x^{■}$

Calculate these powers.

1 6^2

2 4^4

3 $(-2)^3$

4 $(-3)^4$

5 -3^2

6 $-(-5)^3$

7 $\left(\frac{1}{2}\right)^2$

8 0.25^2

Write these numbers as powers.

9 25

10 27

11 -32

12 1000

13 1 000 000

14 0.01

 ISBN: 9780170368186

Fractional powers — roots

- Finding a root is the reverse of finding a power.
- Roots are written as fractional powers.
- $\sqrt{9}$ is the same as writing $\sqrt[2]{9}$ or $9^{\frac{1}{2}}$
- $\sqrt[3]{64}$ is the same as writing $64^{\frac{1}{3}}$

This is asking what **identical** numbers multiply together to get 9.

Examples: $\sqrt{9} = \sqrt{? \times ?} = \sqrt{3 \times 3} = 3$ **or** $\sqrt{9} = \sqrt{? \times ?} = \sqrt{-3 \times -3} = -3$

$\therefore \sqrt{9} = \pm 3$

$\sqrt[3]{-64} = \sqrt[3]{? \times ? \times ?} = \sqrt[3]{-4 \times -4 \times -4}$

$\therefore \sqrt[3]{-64} = -4$

$\sqrt{\text{+ve number}}$ has two answers: $\pm x$.

You can do this on your calculator with either 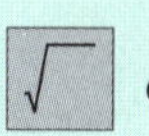 or 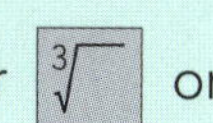 or $\sqrt[x]{\ }$

Calculate these fractional powers.

1 $16^{\frac{1}{2}}$ ______________________

2 $25^{\frac{1}{2}}$ ______________________

3 $125^{\frac{1}{3}}$ ______________________

4 $16^{\frac{1}{4}}$ ______________________

5 $0.25^{\frac{1}{2}}$ ______________________

6 $\left(\frac{1}{4}\right)^{\frac{1}{2}}$ ______________________

Estimating the size of a root

You can estimate the size of a square root by considering the sizes of the perfect squares on either side.

Example: Estimate $\sqrt{30}$.

$5^2 = 25$

$6^2 = 36$

30 lies between 25 and 36

$\therefore \sqrt{30}$ must lie between 5 and 6

Complete the second column **without** using a calculator, then find the matching value in the third column.

Square root	Upper and lower limits for square root	Value
$\sqrt{40}$	6 and 7	8.185
$\sqrt{6}$		4.472
$\sqrt{80}$		6.325
$\sqrt{67}$		2.449
$\sqrt{20}$		8.944

ISBN: 9780170368186

Negative powers

- These are the reciprocal of the positive power.
- To find the reciprocal, you turn a fraction upside down.

Examples: $3^{-2} = \frac{1}{3^2} = \frac{1}{9}$ $(-4)^{-3} = \frac{1}{(-4)^3} = -\frac{1}{64}$

$10^{-2} = \frac{1}{10^2} = \frac{1}{100}$ $\left(\frac{3}{7}\right)^{-2} = \left(\frac{7}{3}\right)^2 = \frac{49}{9}$

Calculate these negative powers.

1 2^{-1}

2 10^{-2}

3 2^{-4}

4 10^{-6}

5 $\left(\frac{5}{4}\right)^{-1}$

6 $\left(\frac{1}{3}\right)^{-1}$

7 $16^{-\frac{1}{2}}$

8 $\left(\frac{1}{9}\right)^{-\frac{1}{2}}$

Mixing it up

1 $3^2 \times 2$

2 3×4^3

3 $-(-4)^2$

4 -4^2

5 $\frac{2^6}{2^2}$

6 $\frac{6^2}{\sqrt{9}}$

7 -5^{-2}

8 $(\sqrt{4})^2$

9 $3^2 + (-3)^2 - (-3)^2$

10 $20 + 5^3$

11 $\sqrt[3]{27} \times 4^2$

12 $(-2 - 4)^2 + (3 \times 2)^2$

ISBN: 9780170368186

Rounding

- Often we need to round numbers to sensible and/or meaningful values.
- Never round until **after** you have completed all calculations.

1 Rounding to decimal places

- When asked to round to 2 dp (two decimal places), this means there should be **exactly** two digits after the decimal point.

Locate the digit to the **right** of the last required decimal place.

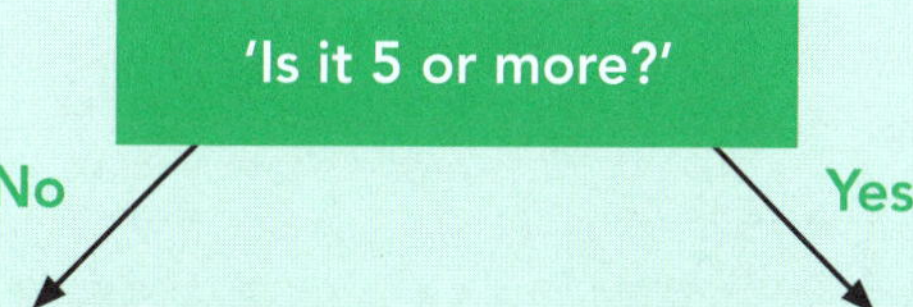

Leave the previous digit alone

Example 1: Round 2.4632 to 2 dp.

The **3** is **not** 5 or more
$\Rightarrow$ leave the 6 alone

$\therefore$ 2.4632 = 2.46 (2 dp)

You are asking whether the original number, 2.4632, is closer to 2.46 or 2.47.

Increase the previous digit by one

Example 2: Round 2.4652 to 2 dp.

The **5 is** 5 or more
$\Rightarrow$ increase the 6 to 7

$\therefore$ 2.4652 = 2.47 (2 dp)

You are asking whether the original number, 2.4652, is closer to 2.46 or 2.47.

Some examples:

Unrounded number	Number of decimal places	Rounded number
0.129	2	0.13
3.824499	3	3.824
1.73051	3	1.731
0.999	1	1.0
4.0962	2	4.10
699.501	0	700

Your answer **must** have the required number of decimal places, so don't omit the 0s.

ISBN: 9780170368186

Round these to the required number of decimal places.

1 6.82331 (2 dp)

2 2.467981 (2 dp)

3 64.1302 (1 dp)

4 142.09 (1 dp)

5 9.0912 (3 dp)

6 91.4458 (3 dp)

7 399.899 (2 dp)

8 0.0314 (2 dp)

9 0.004 (2 dp)

10 1.0203 (2 dp)

11 2.99 (1 dp)

12 2.36^2 (1 dp)

13 $\sqrt{89}$ (2 dp)

14 86.3 x 23.5 (0 dp)

15 $\sqrt[3]{568}$ (1 dp)

16 $\dfrac{4.378 \times 5.699}{\sqrt{76.592}}$ (3 dp)

17 A patient has their temperature taken. It is 38.16 °C.

a Round the temperature to the nearest degree (i.e. 0 dp).

b Round it to the nearest tenth of a degree (i.e. 1 dp).

18 A rugby field is 103.41 m long and 49.26 m wide.

a Calculate its area to the nearest square metre.

b The groundsman only needs to know the area to the nearest 10 m^2 (for fertilising). Round the area to the nearest 10 m^2.

ISBN: 9780170368186

Money

- When dealing with cash in dollars, round to a maximum of 1 dp, i.e. to the nearest 10c.
- Otherwise, round to a maximum of 2 dp, i.e. to the nearest 1c.
- Do **not** round until you reach your final answer.

Example: Aroha bought 850 g of sausages that cost $11.99 per kilo. How much did they cost?

$11.99 x 0.85 = $10.1915
They cost $10.19, or $10.20 if she pays cash.

Calculate the cost of the following.

1 14.5 m of wire at $3.99 per m

2 1½ dozen eggs at $4.75 per dozen

3 1.3 kg of apricots at $6.99 per kg

4 415 g cheese at $29.95 per kg

5 3.1 m of fabric at $19.99 per m

6 345 g fish at $25.99 per kg

7 750 g of potatoes at $2.99 per kg, 450 g of carrots at $1.99 per kg, and a head of broccoli at $1.99

8 650 g of mince at $11.99 per kg and 1.25 kg of rump steak at $18.99 per kg

9 2.5 m of fabric at $24.99 per m, 0.75 m of braid at $7.99 per m, and 1.5 m of ribbon at $2.99 per m

10 Oscar needs 3.5 m of frost cloth at $2.39/m and 7.5 m of bird netting at $4.99/m. Calculate the total price:

a if the two prices are rounded to the nearest cent and then added

b if the two prices are added and then rounded to the nearest cent.

ISBN: 9780170368186

2 Rounding to significant figures

Rules for deciding whether a digit is significant

1 All non-zero digits are significant.

2 Zero **is** significant if it is
- between non-zero digits. **Examples:** 3105 has 4 significant figures; 1.0032 has 5 significant figures
- at the end of a decimal. **Examples:** 2.70 has 3 significant figures; 16.00 has 4 significant figures

3 Zero **is not** significant if it is
- at the start of a number smaller than one. **Examples:** 0.005 has 1 significant figure; 0.592 has 3 significant figures
- at the end of a number which has no decimal point. **Examples:** 2840 has 3 significant figures; 100 has 1 significant figure

Write down the number of significant figures in the following.

Number	Number of significant figures
51.2	
13 706	
14.028	
94 778.5	
10	
3.20046	
4 461 030	
12 000	
1.20	
1205	
0.2	

Number	Number of significant figures
0.004	
43.50	
9.670420	
53.060	
2.0670	
79 000 001	
10 000	
60.0850	
0.060140	
1.030	
0.090	

ISBN: 9780170368186

How to round to significant figures

- When asked to round to 4 sf (four significant figures), this means there should be exactly four significant figures in the answer.
- The process is almost the same as rounding to decimal places.
- **Always** check that the rounded number is a similar size to the original number.

Locate the digit to the **right** of the last required significant figure.

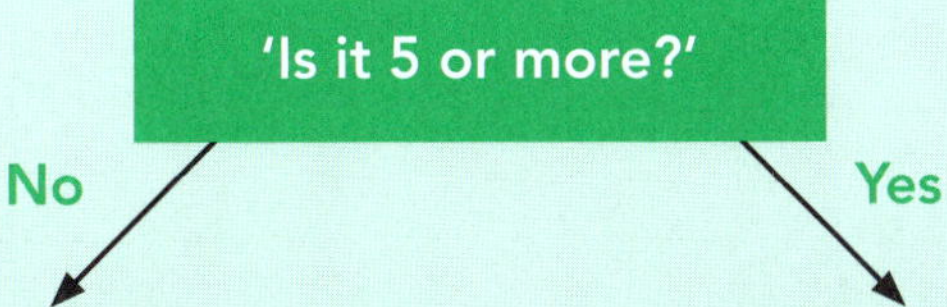

Leave the previous digit alone	Increase the previous digit by one

Example 1: Round 1.50742 to 4 sf.

The **4** is **not** 5 or more
⇒ leave the 7 alone

∴ 1.50742 = 1.507 (4 sf)

You are asking whether the original number, 1.50742, is closer to 1.507 or 1.508.

Example 2: Round 1.50752 to 4 sf.

The **5 is** 5 or more
⇒ increase the 7 to 8

∴ 1.50752 = 1.508 (4 sf)

You are asking whether the original number, 1.50752, is closer to 1.507 or 1.508.

Some examples:

Unrounded number	Number of significant figures	Rounded number
123 456	2	120 000
4.26794	4	4.268
17.500	2	18
0.12079	4	0.1208
0.0207889	4	0.02079
35.7051	4	35.71
29.5	2	30

Sometimes when you round to 2 dp, you get a number that looks as though it is rounded to 1 dp.

Unrounded number	Number of significant figures	Rounded number
67 291	4	67 290
0.999	2	1.0
4.0962	3	4.10
0.0038962	3	0.00390
699.501	1	700
109.9	3	110
1.795	2	1.80

Your answer **must** have the required number of significant figures, so don't omit the 0s.

ISBN: 9780170368186

Round these to the required number of significant figures.

1 52.8 (2 sf)

2 34 651 (2 sf)

3 0.0678 (2 sf)

4 10 462 (2 sf)

5 109.8 (3 sf)

6 0.001055 (3 sf)

7 741 980 (3 sf)

8 0.0069421 (3 sf)

9 1 671 512 (4 sf)

10 0.069995 (4 sf)

11 67.8 (1 sf)

12 94 267 (1 sf)

13 0.47911 (1 sf)

14 0.0555 (1 sf)

15 0.10683 (1 sf)

16 0.00789 (1 sf)

17 $\sqrt{0.887}$ (3 sf)

18 0.125 x 0.863 (3 sf)

19 1.522 ÷ 0.7 (3 sf)

20 0.563 + 0.679 + 0.441 + 0.105 (3 sf)

21 If 11 and 12 have both been rounded to 2 sf, calculate the lowest and highest possible values for the sum of the unrounded numbers.

ISBN: 9780170368186

Estimation

- In order to estimate an answer, round every number to one significant figure, and then do the calculation required.
- Don't forget to use BEDMAS.

Example: $\frac{4.269^2 + 1.832}{3.199} \approx \frac{4^2 + 2}{3}$ ≈ means 'is approximately equal to'.

$= 6$ (The actual answer is 6.270.)

Calculate approximate answers to the following.

1 $1.578 + 5.722 + 9.875 + 0.883 \approx$

2 $\frac{56.91 - 3.674^2}{1.997} \approx$

3 $(5.099 + 1.874)(2.566 - 0.639) \approx$

4 $\frac{1}{1.783}(2.466 + 3.567)^2 \approx$

5 $\frac{9.587 - 4.0367}{3.751 + 2.466} \approx$

6 $\sqrt{8.887} - \frac{9.392 - 0.577}{1.99^2} \approx$

7 $\frac{(3.102)^2 \times \sqrt{98.4}}{1.789} \approx$

8 $\frac{19.65^2}{22.13} \div \frac{\sqrt{9.32}}{0.786} \approx$

9 $(\sqrt[3]{1052})^2 \times \frac{0.435}{2.76 + 7.35} \approx$

10 $(1.68 - 0.84)(0.581 + 4.382)^2 \approx$

11 Tania is going tramping. She wants to get an idea of how heavy her pack will be without actually packing it. The pack itself weighs 2.34 kg, her sleeping bag is 1.84 kg, her clothes will be about 3 kg, and food will be about 2.8 kg. She also needs a billy, knives, plates and a cooker, which weigh 2.96 kg. Her jacket is 1.16 kg. Estimate the total weight of her filled pack.

ISBN: 9780170368186

Fractions

- Fractions are a way of displaying numbers or parts of numbers.
- It can be useful to visualise fractions in the form of a diagram.

Example:

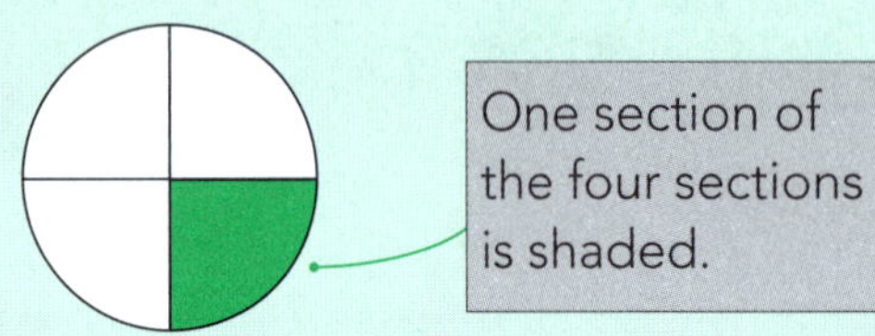

This represents $\frac{1}{4}$.

What fraction of these diagrams is shaded?

1 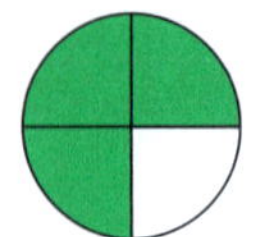 ______

2  ______

3  ______

4 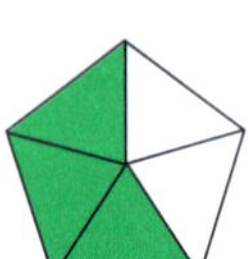 ______

5  ______

6 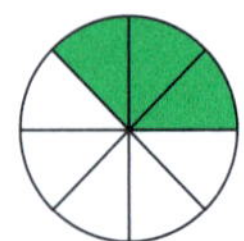 ______

7 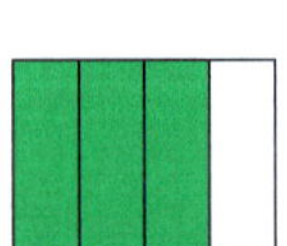 ______

8 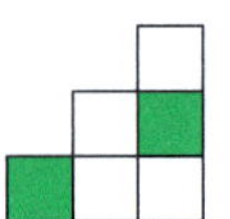 ______

Shade the stated amount in each diagram.

1 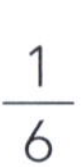 $\frac{1}{6}$ 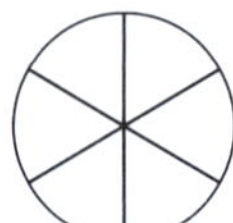

2 $\frac{3}{7}$ 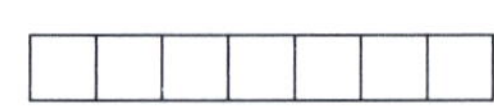

3  $\frac{2}{5}$ 

4 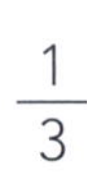 $\frac{1}{3}$ 

5 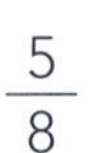 $\frac{5}{8}$ 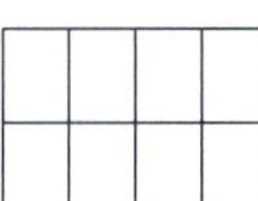

6 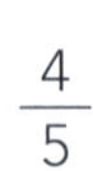 $\frac{4}{5}$ 

7 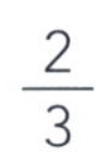 $\frac{2}{3}$ 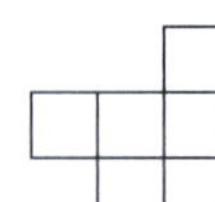

8 $\frac{5}{12}$ 

 ISBN: 9780170368186

Fractions on your calculator

You can put fractions into your calculator with either a b/c or d/c or ▭/▭

5 ⌟ 6 means $\frac{5}{6}$

2 ⌟ 1 ⌟ 3 means $2\frac{1}{3}$

Converting between improper fractions and mixed fractions

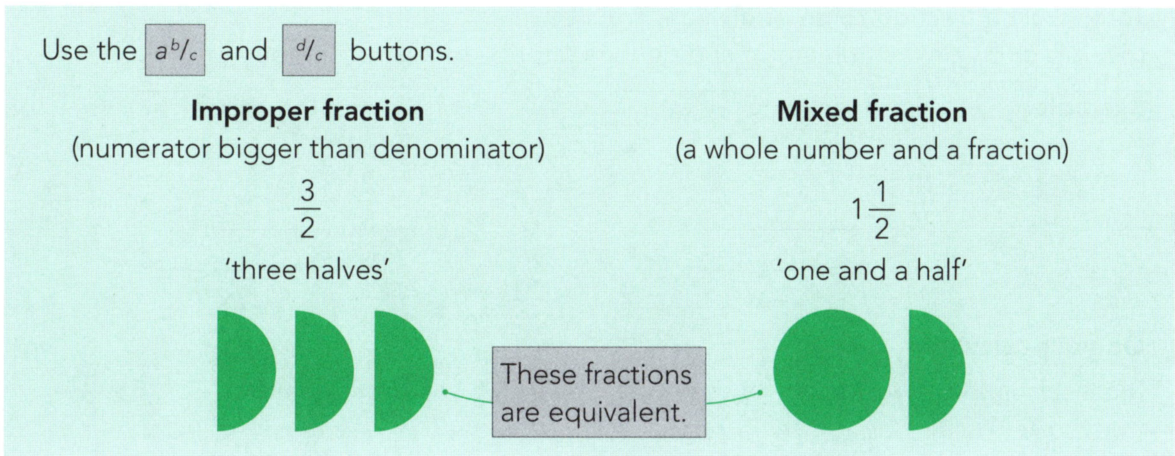

Turn these improper fractions into mixed fractions.

1 $\frac{3}{2}$ ________________

2 $\frac{7}{2}$ ________________

3 $\frac{5}{4}$ ________________

4 $\frac{6}{1}$ ________________

5 $\frac{13}{5}$ ________________

6 $\frac{13}{4}$ ________________

7 $\frac{10}{3}$ ________________

8 $\frac{12}{8}$ ________________

9 $\frac{27}{4}$ ________________

Turn these mixed fractions into improper fractions.

1 $2\frac{1}{2}$ ________________

2 $1\frac{3}{4}$ ________________

3 $3\frac{1}{4}$ ________________

4 $1\frac{1}{9}$ ________________

5 $4\frac{2}{3}$ ________________

6 $5\frac{1}{2}$ ________________

7 $6\frac{9}{10}$ ________________

8 $1\frac{3}{8}$ ________________

9 $7\frac{2}{3}$ ________________

ISBN: 9780170368186

Equivalent fractions

These are fractions that are written differently, but they have the same value.

Example: $\frac{3}{4}$ $\frac{6}{8}$ $\frac{9}{12}$ $\frac{18}{24}$

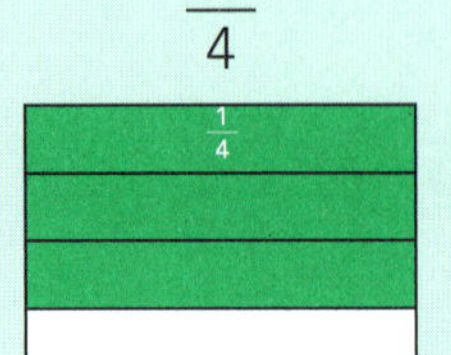

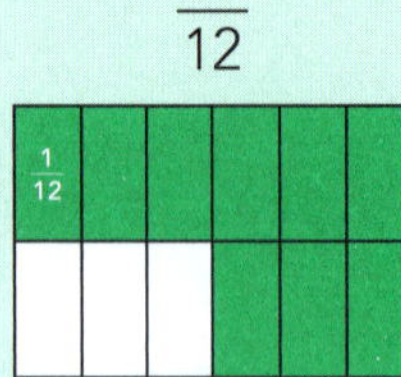

In each case, $\frac{3}{4}$ of the rectangle is shaded. $\frac{3}{4}$ is the simplest form for these fractions.

To convert a fraction to an equivalent fraction

Multiply or divide the numerator and denominator by the **same** number.

Example:

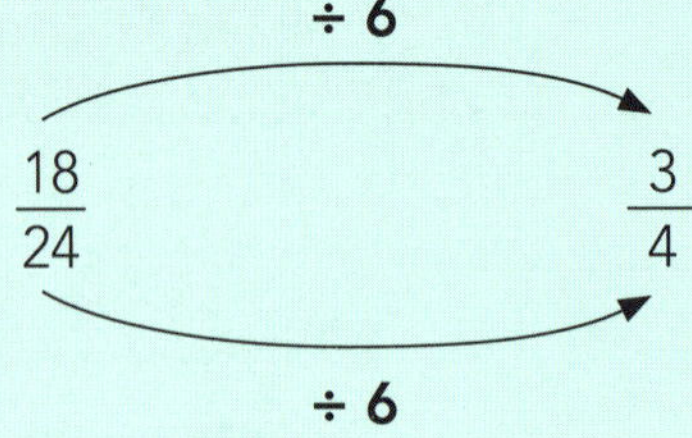

$\frac{2}{5}$ **x 7** → $\frac{14}{35}$ (**x 7**)

On your calculator

Most calculators, if you enter, for example, 18 $a^b/_c$ 24, will display either $\frac{3}{4}$ or 3⌟4.

Write each fraction in its simplest form.

1 $\frac{3}{9}$ ______

2 $\frac{4}{12}$ ______

3 $\frac{15}{20}$ ______

4 $\frac{90}{100}$ ______

5 $\frac{21}{35}$ ______

6 $\frac{18}{20}$ ______

7 $\frac{250}{1000}$ ______

8 $\frac{27}{54}$ ______

9 $\frac{36}{48}$ ______

10 $\frac{35}{60}$ ______

Fill in the gaps to create equivalent fractions.

1 $\frac{2}{3} = \frac{}{48}$

2 $\frac{13}{117} = \frac{1}{}$

3 $\frac{3}{7} = \frac{45}{}$

4 $\frac{100}{125} = \frac{4}{}$

5 $\frac{70}{84} = \frac{}{6}$

6 $\frac{3}{8} = \frac{}{104}$

7 $\frac{55}{132} = \frac{5}{}$

8 $\frac{60}{165} = \frac{4}{}$

9 $\frac{34}{51} = \frac{}{3}$

10 $\frac{7}{10} = \frac{84}{}$

ISBN: 9780170368186

Calculations involving fractions

You can use a number of functions on your calculator to add, subtract, multiply and divide fractions.

$\frac{1}{2} + \frac{1}{3} = \frac{5}{6}$

$1\frac{1}{3} - \frac{1}{4} = 1\frac{1}{12}$ or $\frac{13}{12}$

Answers could be mixed or improper.

$\frac{4}{5} \times \frac{2}{5} = \frac{8}{25}$

$\frac{4}{5} \div \frac{1}{3} = 2\frac{2}{5}$ or $\frac{12}{5}$

Add these fractions.

1a $\frac{1}{2} + \frac{3}{4} =$ ____________________

b $\frac{4}{5} + \frac{2}{3} =$ ____________________

c $\frac{1}{8} + \frac{2}{5} =$ ____________________

d $\frac{7}{8} + \frac{1}{4} =$ ____________________

e $\frac{3}{5} + 2\frac{3}{4} =$ ____________________

f $1\frac{3}{7} + \frac{5}{1} =$ ____________________

g $\frac{9}{3} + \frac{1}{2} =$ ____________________

h $\frac{2}{1} + \frac{3}{3} =$ ____________________

i $1\frac{4}{5} + \frac{1}{4} =$ ____________________

j $\frac{1}{7} + 4\frac{7}{8} =$ ____________________

Subtract these fractions.

2a $\frac{1}{2} - \frac{1}{4} =$ ____________________

b $\frac{2}{3} - \frac{1}{5} =$ ____________________

c $\frac{7}{8} - \frac{3}{4} =$ ____________________

d $\frac{3}{5} - \frac{1}{3} =$ ____________________

e $2\frac{1}{2} - \frac{2}{3} =$ ____________________

f $1\frac{1}{4} - \frac{8}{9} =$ ____________________

g $2\frac{1}{6} - \frac{3}{4} =$ ____________________

h $1\frac{4}{5} - 1\frac{1}{4} =$ ____________________

i $3\frac{2}{5} - \frac{1}{4} - \frac{1}{3} =$ ____________________

j $1\frac{1}{19} - \frac{1}{20} =$ ____________________

ISBN: 9780170368186

Multiply these fractions.

3a $\frac{1}{5} \times \frac{2}{3} =$ ______

b $\frac{2}{5} \times \frac{1}{4} =$ ______

c $\frac{7}{8} \times \frac{2}{5} =$ ______

d $\frac{1}{6} \times \frac{6}{7} =$ ______

e $\frac{7}{5} \times \frac{1}{3} =$ ______

f $1\frac{1}{4} \times \frac{2}{3} =$ ______

g $2\frac{2}{3} \times \frac{1}{3} =$ ______

h $1\frac{2}{5} \times 2\frac{2}{6} =$ ______

i $2\frac{3}{4} \times 7 =$ ______

j $\frac{2}{3} \times \frac{1}{3} \times \frac{1}{4} =$ ______

Divide these fractions.

4a $\frac{1}{5} \div \frac{2}{3} =$ ______

b $\frac{7}{8} \div \frac{1}{4} =$ ______

c $\frac{2}{5} \div \frac{1}{3} =$ ______

d $\frac{7}{5} \div \frac{1}{5} =$ ______

e $1\frac{1}{2} \div \frac{1}{4} =$ ______

f $2\frac{2}{3} \div \frac{1}{4} =$ ______

g $3\frac{4}{5} \div 1\frac{2}{3} =$ ______

h $\frac{3}{4} \div 1\frac{2}{3} =$ ______

i $6 \div \frac{1}{5} =$ ______

j $\frac{1}{5} \div 6 =$ ______

5a Dave, Divesh, Delia and Dan order a pizza. Dave eats $\frac{5}{12}$, Divesh eats $\frac{1}{4}$, and the rest is shared equally between Delia and Dan. What fraction of the pizza did Delia eat?

______

______

b If Dave had eaten only $\frac{1}{3}$ of the pizza, instead of $\frac{5}{12}$, how much pizza would Delia have eaten?

______

______

 ISBN: 9780170368186

Fractions and recurring decimals

- Recurring decimals are decimals in which a pattern of numbers is repeated.
- Do **not** ignore the dot, which means a decimal is recurring — it can make a large difference.

Example: 0.6 of \$1000 = \$600
$0.\dot{6}$ of \$1000 = \$666.67 — that's a difference of almost \$67!

How to write recurring decimals

Fraction	Decimal	Notation
$\frac{1}{9}$	0.11111111111…	$0.\dot{1}$
$\frac{3}{11}$	0.27272727272…	$0.\dot{2}\dot{7}$
$\frac{7}{15}$	0.466666666666…	$0.4\dot{6}$

Fraction	Decimal	Notation
$\frac{9}{44}$	0.204545454545…	$0.20\dot{4}\dot{5}$
$\frac{11}{27}$	0.407407407407…	$0.\dot{4}0\dot{7}$
$\frac{1}{7}$	0.142857142857…	$0.\dot{1}4285\dot{7}$

Write the following decimals in full.

1 $0.\dot{6}$ ______

2 $0.\dot{4}\dot{7}$ ______

3 $0.5\dot{9}\dot{1}$ ______

4 $0.\dot{4}1\dot{6}$ ______

5 $5.11\dot{2}\dot{4}$ ______

6 $10.13\dot{6}$ ______

7 $10.13\dot{6}\dot{3}$ ______

8 $10.13\dot{6}3\dot{6}$ ______

Write the following decimals using recurring decimal notation.

9 0.77777777… ______

10 0.45454545… ______

11 9.14141414… ______

12 2.37575757… ______

13 20.02323232… ______

14 9.54654654… ______

15 1.00100100… ______

16 0.23642364… ______

Write the following fractions as decimals in recurring decimal notation.

17 $\frac{2}{3}$ ______

18 $\frac{5}{11}$ ______

19 $\frac{5}{6}$ ______

20 $\frac{6}{55}$ ______

21 $\frac{8}{27}$ ______

22 $\frac{7}{44}$ ______

23 $\frac{9}{555}$ ______

24 $\frac{60}{999}$ ______

ISBN: 9780170368186

Deciding which of two fractions is larger

The easiest way is to use your calculator to convert both fractions to decimals.

Example: Which is larger, $\frac{5}{8}$ or $\frac{8}{13}$?

Is greater than

$\frac{5}{8} = 0.625$ $\frac{8}{13} = 0.6154$ (4 dp) $0.625 > 0.6154$ $\therefore \frac{5}{8}$ is larger

Convert the following fractions to decimals and highlight the larger one.

1 $\frac{1}{3}$ = ________ $\frac{3}{10}$ = ________ **2** $\frac{1}{4}$ = ________ $\frac{4}{15}$ = ________

3 $\frac{4}{7}$ = ________ $\frac{6}{11}$ = ________ **4** $\frac{2}{3}$ = ________ $\frac{8}{13}$ = ________

5 $\frac{2}{5}$ = ________ $\frac{5}{13}$ = ________ **6** $\frac{19}{24}$ = ________ $\frac{3}{4}$ = ________

7 $\frac{5}{6}$ = ________ $\frac{17}{20}$ = ________ **8** $\frac{10}{11}$ = ________ $\frac{9}{10}$ = ________

Answer the following questions, and justify your answers.

9 Dan got 20 questions correct out of 30 in his science test, and 7 questions correct out of 11 in his maths test. In which subject did he do better?

____

____

10 Avocados were five for $2.00 at a market stall, and seven for $3.00 at the supermarket. Where were they cheapest?

____

____

11 The local cinema sells books of five tickets for $49.95. The city cinema sells books of eight tickets for $75. At which cinema are the tickets cheapest?

____

____

12 A 1.5 kg bag of flour costs $2.79, and a 5 kg bag costs $10.99. Which bag contains cheaper flour?

____

____

ISBN: 9780170368186

Fractions of an amount

Find $\frac{2}{3}$ **of** 54: $\frac{2}{3}$ **of** $54 = \frac{2}{3}$ **x** 54

$= 36$

'**of**' means multiply

Find $\frac{1}{5}$ of \$95: $\frac{1}{5}$ of $\$95 = \frac{1}{5} \times \$95$

= **\$**19

Don't forget the **units**

Answer the following questions.

1 $\frac{1}{3}$ of 156 kg

2 $\frac{4}{5}$ of 315 mL

3 $\frac{1}{2}$ of \$89

4 $\frac{2}{7}$ of 133 km

5 $\frac{1}{4}$ of 96 m

6 $1\frac{2}{3}$ of 177 cm

7 $2\frac{1}{5}$ of 46.5 g

8 $\frac{8}{3}$ of \$93

9 $\frac{3}{10}$ of 236 kg

10 $\frac{2}{5}$ of 85%

11 $\frac{5}{4}$ of 252.8 L

12 $\frac{7}{7}$ of 58 km

13 $\frac{1}{3}$ of 1704 cm

14 $\frac{3}{4}$ of $\frac{1}{2}$

15 $\frac{1}{3}$ of 0.26

16 Five students drive to Dunedin, and share the cost of the fuel. If the fuel cost \$52.90, how much did each student pay?

17 Two families, one of three people and one of five people, hire a bach for \$196 per night. If the cost is shared equally among the eight people, how much will it cost the family of three per night?

18 Scott ate $\frac{1}{6}$ of a cake. Mel then ate $\frac{3}{10}$ of what was left. What fraction of the entire cake did Mel eat?

19 Uncle Scrooge's will states that one third of his estate goes to the cat. The remainder is to be shared equally between his five nieces. What fraction of his estate does each niece inherit?

20 Pania ate a quarter of the pizza. Pita ate one sixth of what remained. Paula ate two fifths of what remained after Pita had eaten. What fraction of the pizza was left for Penny?

ISBN: 9780170368186

Applications of fractions

Remember that all of something is one whole.

Answer the following questions.

1 Pip's mud pie contains 2 cups of water for every 3 cups of dry soil. What fraction of the pie is water?

2 A fruit bowl contains 5 apples, 3 plums and 7 apricots. What fraction of the pieces of fruit are apricots?

3 Mick spent $\frac{1}{3}$ of his pocket money on a movie ticket, and $\frac{1}{8}$ on bus fare. What fraction of his pocket money was left?

4 A recipe for fruit cake lists $1\frac{1}{3}$ cups of sultanas and $\frac{3}{4}$ cup of raisins. How many cups of dried fruit is this?

5 Oliver ate 3 slices of cake and Ian ate 4 slices. If there were 5 slices remaining, what fraction of the cake did they eat?

6 Ngaire ran $2\frac{3}{4}$ km on Saturday and $1\frac{7}{8}$ km on Sunday. How much further did she run on Saturday than on Sunday?

7 Nick skateboards to school half the time, he goes on the bus a third of the time, and his mother drops him off on the remaining days. On what fraction of school days does his mother drop him at school?

8 Airini, Sam and Jess painted their Grandma's fence. If Airini painted for $2\frac{1}{2}$ hours, Sam for $1\frac{3}{4}$ hours, and Jess for $1\frac{1}{6}$ hours, how many hours did it take altogether?

9 Eighteen students share $4\frac{1}{2}$ kg of cherries equally. What mass of cherries does each student get?

10 Five sixths of Adam's lollies are jelly beans. If he swaps two thirds of his jelly beans for a marble, what fraction of his lollies did the marble cost?

11 Anna was sick for five out of the 40 mathematics periods in the term, and she had music lessons during three periods. For what fraction of the mathematics periods was she present?

12 Zoe's pack contains a sleeping bag, which weighs $3\frac{1}{2}$ kg, her clothes, $2\frac{3}{4}$ kg, and her food, $2\frac{1}{3}$ kg. If her filled pack weighs $10\frac{1}{2}$ kg, what does her empty pack weigh?

ISBN: 9780170368186

Mini-task

A class of 24 students is planning a picnic. They need to buy:

- pizzas at $7.99 each, and each pizza is to be shared between five people
- an ice block for each person, which cost $7.49 for a mixed packet of four
- a carton of juice for each person; these cost $1.99 each
- 2.15 kg mandarins at $4.49 per kilo
- lollies for the treasure hunt: a packet of butterscotch and a packet of mints, which each cost $3.99, and a packet of toffees, which cost $4.99.

In the space below, work out the following.

- The total cost of the food.
- The total cost for each student.
- The ice blocks were stored in a school freezer, but Adam left the door open. The freezer is supposed to be at -4 °C, but by the time it was discovered, the temperature was 1 °C. By how much had the temperature risen?
- $\frac{1}{4}$ of the ice blocks were raspberry, $\frac{1}{3}$ were lemonade, and the rest were orange. What fraction were orange?
- The lollies are to be divided up into identical smaller bags, each of which must have the same number of each flavour. There will not be enough bags for everybody — they are to be used for a treasure hunt. There are 24 butterscotch, 30 mints and 42 toffees. What is the maximum number of bags that they can make, and how many of each type of lolly will be in each one?

ISBN: 9780170368186

Converting between decimals, fractions and percentages

Fractions are a way of displaying parts of numbers. You can convert fractions into decimals using your calculator with either $a^b/_c$ or ⊟ or S⟺D

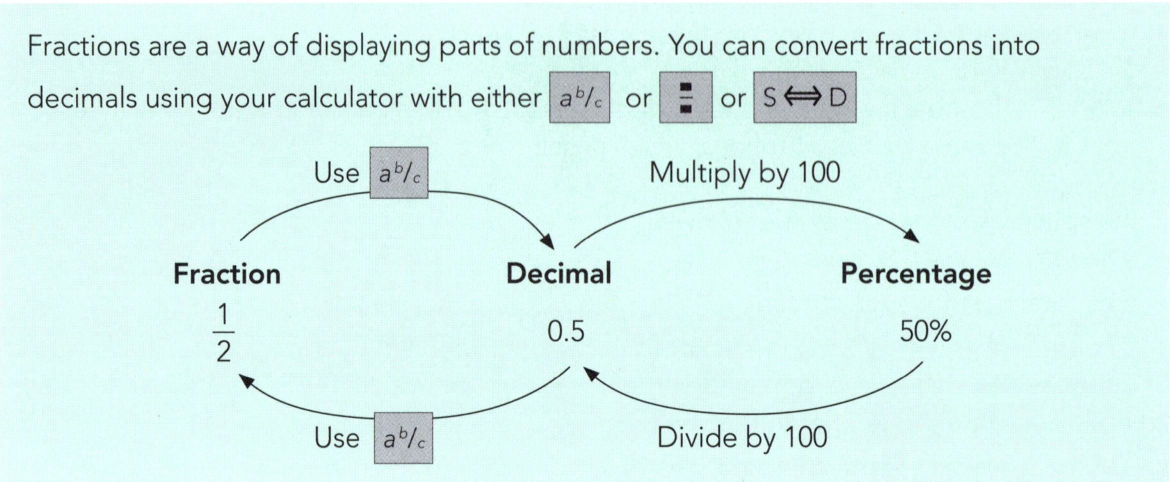

Fill in the blanks in the table below.

Fraction	Decimal	Percentage
$\frac{1}{4}$		
	0.75	
		30%
	0.2	
$\frac{3}{5}$		
		68%
	0.12	
$\frac{7}{8}$		
		100%
	2.5	
		4%
	1.2	

  ISBN: 9780170368186

Percentages

Calculating percentages

- To turn a fraction into a percentage, multiply by 100.

Example: 47 out of 64 Year 11 girls play sport. What percentage play sport?

$\frac{47}{64} \times 100 = 73.4\%$ (1 dp) or use the **percentage** button on your calculator: $\frac{47}{64}$ [%] = 73.4% (1 dp)

Convert these into percentages.

1 There are 14 girls in a class of 25. What percentage are girls?

2 It rained on eight days last April. On what percentage of days did it rain?

3 Nine out of the 20 pieces of fruit in a bowl were bananas. What percentage were bananas?

4 Annie was absent from five out of 42 mathematics lessons. For what percentage was she present?

5 The normal price of a pair of jeans was $85. In a sale, $15 was taken off this price. By what percentage were they reduced?

6 A muesli recipe requires 6 cups of dry ingredients, 1 cup of which is bran. What percentage of the dry ingredients is bran?

7 Tamara earned 108 credits in NCEA Level 1. Of these, 21 were excellence credits, 53 were merit, 28 were achieved and 6 were not achieved. Calculate the percentage of each type of credit.

8 Eru was doing a survey of vehicles entering the school. Of the 38 vehicles, 22 were private cars, 9 were motor bikes, 4 were taxis and 3 were delivery vans. Calculate the percentage of each type of vehicle.

9 When Amelia was in Australia, her bank charged her $5 per transaction on her eftpos card. Calculate the percentage of the **total** transaction that she pays the bank when

a she buys a $35 shirt using eftpos.

b she withdraws $200 cash using her eftpos card.

ISBN: 9780170368186

Percentages of an amount

Example:

Find 10% **of** 84

'**of**' means multiply.

Use the **%** button on your calculator.

10% of 84 = **0.10** x 84
= 8.4

or

10% of 84 = 84 x 10**%**
= 8.4

Convert the percentage to a **decimal**.

Calculate these.

1 20% of 265 km

2 50% of 248 mm

3 25% of $522

4 75% of 368 mL

5 45% of 68 g

6 2% of 78 kg

7 14% of 183 cm

8 200% of $7.50

9 120% of $400

10 16% of 48 mL

11 10.5% of $126

12 0.5% of $826

13 1.5% of 672 km

14 3.5% of $5000

15 0.45% of $1.2 million

16 A football team won 75% of 60 games in a season. How many games did it win?

17 Luke and Matthew ran a lemonade stand on Saturday. They agreed that Matthew would get 60% of the profit because the lemonade stand was his idea. They made a profit of $25. How much money did Matthew make?

18 In a school, 25% of 312 students bring lunch from home. How many students bring lunch from home?

19 Uncle Scrooge revised his will. The new one states that 40% goes to the cat. The remainder is to be shared equally between his five nieces. If his estate is worth $38 000, calculate how much the cat inherits, and how much each niece inherits.

 ISBN: 9780170368186

Increasing by a percentage

Example:

Increase 70 by 10%.

Means we need 100% plus 10%.

Increased amount = 70 x **1.1**
= 77

Convert the percentage to a **decimal and add 1**.

or

Use the **%** button on your calculator.

Increased amount = 70 + 10**%**
= 77

Calculate these.

1 Increase $45 by 20%.

2 Increase 85 km by 50%.

3 Increase 580 cm by 36%.

4 Increase 0.2 mm by 25%.

5 Increase 512 m by 80%.

6 Increase 48 g by 46%.

7 Increase 9.3 g by 98%.

8 Increase 450 mL by 35%.

9 Increase 87 km by 33.3%.

10 Increase 2 min by 65%.

11 Increase 0.03 kg by 70%.

12 Increase 85 g by 120%.

13 Increase 48 km by 200%.

14 Increase 6.6 kg by 350%.

15 Increase 80 by 0.2%.

16 Alex has $4560 in the bank. He is paid interest of 3.5% at the end of the year. How much does he have in his account at the start of the next year?

17 If a person is between one and seven days late paying tax, they are charged an extra 1% on the amount owed.
If a person is over seven days late paying tax, they are charged an extra 4% on the **total** amount owed (including the extra 1% for being one to seven days late).

a Calculate how much Teone must pay if he owes $380 and he pays it three days late.

b Calculate how much Amelia must pay if she owes $2375 and she pays it 10 days late.

Decreasing by a percentage

Example:

Decrease 60 by 10%.

Means we need 100% minus 10%.

Use the % button on your calculator.

Decreased amount = 60 x **0.9**
= 54

or

Decreased amount = 60 – 10%
= 54

Convert the percentage to a **decimal and subtract from 1**.

Calculate these.

1 Decrease $105 by 30%.

2 Decrease 280 m by 15%.

3 Decrease 195 mm by 50%.

4 Decrease 48 cm by 16%.

5 Decrease 1.6 L by 10%.

6 Decrease 14.4 kg by 25%.

7 Decrease 55 cm by 95%.

8 Decrease 192 L by 7.5%.

9 Decrease 540 g by 5.5%.

10 Decrease 48 min by 8%.

11 Decrease 1.4 tonnes by 9%.

12 Decrease 88 g by 95%.

13 Decrease 92 km by 100%.

14 Decrease 78 km by 2.5%.

15 Decrease $46 by 0.5%.

16 A website advertises '35% discount on all items'. Calculate the discounted price of:

a a phone which is normally $599.

b a hard drive which is normally $145.

17 Sai borrowed $390 from his mum to buy a phone. His mum says she will reduce his debt by 5% of what he owes at the end of each month, provided he keeps his room tidy. If he manages to keep his room tidy for three months, what does he owe her at the end of that period?

ISBN: 9780170368186

Finding a percentage increase or decrease

Example 1: Huia's pay is increased from \$14.50 to \$14.79 per hour. Calculate the percentage increase.

Step 1: **Subtract** the amounts: \$14.79 – \$14.50 = \$0.29

Step 2: **Divide** the difference by the **original** quantity and multiply by 100: $\frac{0.29}{14.50} \times 100 = 2\%$

Example 2: She bought a shirt in a sale. The original price was \$40, but it was reduced to \$31.20. By what percentage was it reduced?

Step 1: **Subtract** the amounts: \$40.00 – \$31.20 = \$8.80

Step 2: **Divide** the difference by the **original** quantity and multiply by 100: $\frac{8.80}{40.00} \times 100 = 22\%$

Calculate the percentage changes below.

1 From 18 to 18.9

2 From 1200 to 1080

3 From 0.480 to 0.552

4 From \$36.00 to \$25.20

5 From 720 to 540

6 From 15 kg to 15.3 kg

7 From 7600 to 7524

8 From 12 to 27

9 From 36 to 261

10 From 135 to 18.9

11 Louisa bought a car for \$3600, but had to sell it six months later. She got \$2736. Calculate her percentage loss on the car.

12 The population of an island increased from 15 864 to 17 843. Calculate the percentage increase in the population.

ISBN: 9780170368186

GST

- **GST** stands for **G**oods and **S**ervices **T**ax.
- It is added to everything you purchase and it goes to the Government to fund the running of the country.
- The standard current GST rate on all products in New Zealand is 15%.
- Don't forget to add units and round when necessary.

Adding GST

Example:
The GST-exclusive price is $95. Calculate the GST-inclusive price.

Use the **%** button on your calculator.

GST-inclusive amount = $95 x **1.15**
= $109.25

Convert the percentage to a **decimal and add 1**.

or

GST-inclusive amount = $95 + 15**%**
= $109.25

Add GST to these prices.

1 $86

2 $15.50

3 $89.95

4 $102.32

5 $4056.99

6 $56.75

7 $0.23

8 $65.20

9 $530.00

10 The pre-GST price of a car is $3500.

a Calculate its GST-inclusive price.

b How much is the GST on the car?

11 The pre-GST price of an ice-cream is $3.47.

a Calculate its GST-inclusive price.

b How much is the GST on the ice-cream?

ISBN: 9780170368186

Calculating the pre-GST price and GST

You may be given the GST-inclusive price and asked to calculate a pre-GST price or how much GST is being paid on an item.

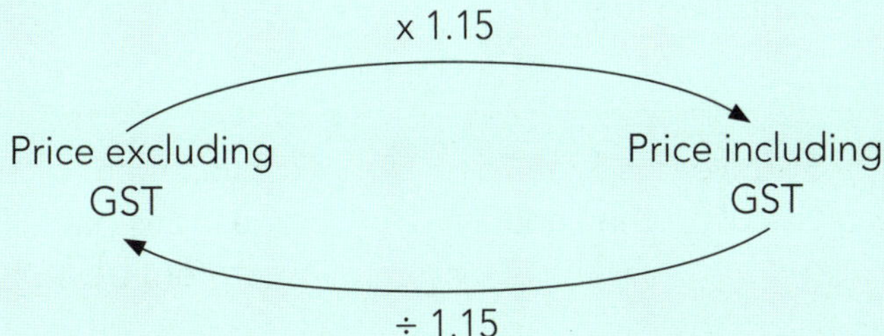

Example: A painting costs \$840 including GST. How much GST was paid?

GST-exclusive price = \$840 ÷ 1.15
= \$730.43 (2 dp)

This is the price without GST.

∴ GST = \$840 – \$730.43
= \$109.57

Calculate GST-exclusive prices for the following.

1 \$98

2 \$27.45

3 \$109.99

4 \$256.78

5 \$0.99

6 \$980.65

7 \$1.59

8 \$86.70

9 \$890.00

10 Derek bought a car for \$4999. How much GST was included in this price?

11 How much GST did Amanda pay on her ice cream that cost her \$3.45?

12 A car yard had a sale. It advertised that it would pay the GST on all cars sold. The salesman subtracted 15% from the full GST-inclusive price of \$5999 in order to calculate the sale price. Explain why his boss was angry, and calculate what the sale price should have been.

ISBN: 9780170368186

Mixing it up

Answer the following questions.

1 Two ninths of the class of 27 caught the flu. How many students had the flu?

2 Calculate the GST-exclusive price of an item that is advertised at $1099.

3 An investment of $5750 earns $235.75 interest in a year. Calculate the annual interest rate.

4 An investment of $2500 earns 3.5% interest per year. How much will the investment be worth after one year?

5 Nic ran 2.6 km on Monday, 3.2 km on Wednesday, 2.9 km on Friday and 4.8 km on Sunday. Calculate the total distance that he ran over the week.

6 A dressing recipe contains four tablespoons of oil, one tablespoon of vinegar and one tablespoon of lemon juice. What percentage of the dressing is lemon juice?

7 A family of three and a family of four go out to dinner. The total bill is $287.35. Calculate how much the family of three should pay.

8 A bike shop has a sale in which 15% is taken off all prices. Calculate the sale price of a bike that is normally $399.

9 The dry ingredients for a muesli bar recipe are $2\frac{1}{2}$ cups of rolled oats, $\frac{3}{4}$ cup of raisins, $1\frac{3}{4}$ cups of nuts and $\frac{1}{2}$ cup of bran. How many cups of dry ingredients is this?

10 Jon got five sixths of the questions in his science test correct, he got 82% in his mathematics test, and 19 questions correct out of 23 in history. In which subject did he score the highest mark?

11 Pita ate a sixth of the pizza. Pania ate three tenths of what remained. Then Paula ate a quarter of the entire pizza. What fraction was left for Penny?

12 An old model of a tablet is advertised at $399, rather than the normal price of $469. By what percentage was the tablet reduced?

13 After a 5% increase in the cost of a litre of milk, the price is $2.73. What was the price before the increase?

14 The price of wine gums increased from 5c each to 6c. Calculate the percentage increase in price.

15 Uncle Scrooge revised his will yet again. The new one states that 35% goes to the cat. The remainder is to be shared equally between his five nieces, except for the youngest (his favourite), who gets an extra 5% of the estate. If his estate is worth $38 000, calculate how much the cat inherits, and how much each niece inherits.

ISBN: 9780170368186

Interest

- When you deposit money in the bank, the bank pays you 'interest' for the use of that money.
- When you borrow money, you have to pay 'interest' because you are using the bank's money.
- Interest may be simple interest (SI) or compound interest (CI).

Simple interest (SI)

- The same amount of interest is paid to you at regular intervals.
- This interest is paid separately into another account: it is **not** added to your deposit.

Example: Rawiri's grandmother has lent him $4500 to buy a car. Rawiri will pay her 5% of $4500 in interest each year.

a How much does he pay each year?

Interest = 5% of $4500
= $225

b If he pays all the money back after three years, how much interest will he have paid to his grandmother in total?

Interest = 3 x $225
= $675.00

Answer the following questions. All interest rates are annual.

1 Calculate the simple interest earned when $1500 is invested at 7% for five years.

____

2 Calculate the simple interest earned when $55 000 is invested at 4.5% for 10 years.

____

3 Calculate the simple interest earned when $8750 is invested at 8% for two years.

____

4 Calculate the rate of interest needed in order to earn $66 per year from a $1200 loan.

____

5 Calculate the rate of interest needed in order to earn $1562.50 per year from a $25 000 loan.

____

6 How long will $4000 need to be invested at 6.5% in order to earn $1040 in interest?

____

ISBN: 9780170368186

Compound interest (CI)

- With compound interest, the interest is added to the amount deposited.
- The following year, interest is calculated on the total (deposit and previous year's interest).

Example: Rawiri's grandmother lends him $4500 to buy a car, and she charges him 5% per year compound interest. This means he doesn't pay her anything annually, but pays back the $4500 plus compound interest after three years.

a How much would he pay back at the end of the three years?

At the end of the 1st year he owes $4500 + $4500 x 5% = $4725
At the end of the 2nd year he owes $4725 + $4725 x 5% = $4961.25
At the end of the 3rd year he owes $4961.25 + $4961.25 x 5% = $5209.31

You can do this on your calculator by entering
$4500 x 1.05 x 1.05 x 1.05 = or $4500 x $(1.05)^3$ =

b Compare this with the amount he would pay his grandmother when the money earned simple interest (page 45).

Simple interest: He would pay his grandmother $4500 + $675 interest = $5175.
Compound interest: He would pay his grandmother $5209.31.
∴ he would pay her $34.31 more if she charged him compound interest.

Answer the following questions.

1 $7500 is invested at 6% compound interest. Calculate its value at the end of two years.

____

2 $25 000 is invested at 6.5% compound interest. Calculate its value at the end of three years.

____

3 What is the value of $65 000 invested at 4.5% compound interest at the end of 10 years?

____

4 **a** When Jane was born, her grandfather invested $2000 at an interest rate of 4.5%. Jane cannot use this until she is 18. How much will it be worth by then?

____

b How much more would it have been worth if it had been invested at 5%?

____

ISBN: 9780170368186

Mini-task

Angus was left $2000 by his grandfather, and the money is deposited in a non-interest-bearing account. His mum says she will add 5% of the total in the bank at the end of each year as long as he doesn't withdraw any of it (compound interest). His father says he has a better offer. If Angus withdraws the money after one year, he will add 5%. If he withdraws it after two years he will add 10%, after three years 15%, etc. Angus has to choose between his mother's and his father's offers.

- Complete the table below to show how much Angus would have after one, two, three or four years.

Money withdrawn at the end of	Mum's scheme ($)	Dad's scheme ($)
one year		
two years		
three years		
four years		

- Which offer would you recommend that Angus accepts? Explain your answer.

Angus accepts his dad's offer, and withdraws the money after four years.

- He uses some to buy a good watch. The normal price is $699, but he buys it when there is a '25% off' sale.
- He also buys a new cricket bat (his grandfather loved cricket too), which costs him $399 plus GST.
- He buys a tablet. Its normal GST-inclusive price is $499 but the shop has a 'we pay the GST' sale.
- He keeps $500 aside to pay for a sports trip.

Use the space below to calculate how much of his money he will have left to contribute to his student fees.

ISBN: 9780170368186

Ratios

Simplifying ratios

- Ratios show how an amount is split into several shares, often of different sizes.
- You should give ratio answers as whole numbers, never decimals or fractions.
- A **colon** (:) is used to separate the parts, e.g 2:1 means 'two **to** one'.
- Like fractions, ratios should be written in their simplest form.
- Ratios do not have units.

Note: Highest common factor (HCF) is the biggest number that will divide into several numbers. For example, the HCF of 15 and 12 is 3.

Example 1: Write the ratio 60:36:24 in its simplest form.

$$60:36:24 = 5:3:2$$

Divide each of these numbers by their HCF (**12**).

These numbers have no common factors.

Example 2: Write the ratio $1\frac{1}{2}:\frac{2}{3}:2$ in its simplest form.

$$1\frac{1}{2}:\frac{2}{3}:2 = 1\frac{1}{2} \times \mathbf{6} : \frac{2}{3} \times \mathbf{6} : 2 \times \mathbf{6} = 9:4:12$$

Multiply each of these numbers by a number big enough to get rid of the fractions (**6**).

Example 3: Write the ratio 0.5 : 0.15 : 0.2 in its simplest form.

$$0.5 : 0.15 : 0.2 = 50:15:20 = 10:3:4$$

Multiply each of these numbers by a number big enough to get rid of the decimal point (**100**).

Then divide each number by its HCF (**5**).

Write the following ratios in their simplest forms.

1 6:4

____

2 100:70:30

____

3 16:20

____

4 $\frac{1}{2}:\frac{3}{4}$

____

5 $3\frac{1}{2}:2\frac{1}{2}$

____

6 $\frac{1}{2}:\frac{1}{3}:\frac{1}{4}$

____

 ISBN: 9780170368186

7 0.1:0.5

8 1.2:3.6:2.4

9 0.04:0.06:0.12

10 12 m : 0.5 m

11 1 hour : 10 min

12 2 km : 3.5 km : $\frac{1}{2}$ km

Ratio calculations where the total amount is given

Example 1: $45 needs to be split between two people in the ratio 2:3.

Step 1: Calculate the **total number of parts** by adding the numbers in each share: 2 + 3 = **5**

Step 2: Calculate the **value of each part** by dividing the total amount by the number of parts: $45 ÷ 5 = **$9**

Step 3: Calculate the **value of each share** by multiplying the number of parts by the value of each part: 2 x $9 : 3 x $9
∴ $18 : $27

One person gets $18 and the other gets $27. (Check: $18 + $27 = $45)

Example 2: Annie, Amira and Amelia paint their grandma's fence. She will pay them a total of $180. Annie works for five hours, Amira for four hours, but Amelia gets bored and leaves after three hours. How much should each girl get paid?

Step 1: Calculate the **total number of parts** by adding the numbers in each share: 5 + 4 + 3 = **12**

Step 2: Calculate the **value of each part** by dividing the total amount by the number of parts: $180 ÷ 12 = **$15**

Step 3: Calculate the **value of each share** by multiplying the number of parts by the value of each part: 5 x $15 : 4 x $15 : 3 x $15
∴ $75 : $60 : $45

Annie gets $75, Amira gets $60 and Amelia gets $45.
(Check: $75 + $60 + $45 = $180)

Solve the following ratio problems.

1 Sam and Sarah pool their money to buy a $3 bag of lollies. Sam gave $2 and Sarah gave $1. If there are 51 lollies in the bag, how many lollies should each get?

2 A recipe requires two parts of liquid for every three parts of dry ingredients. If Hazel wants to make a total of $2\frac{1}{2}$ cups, how much of each type of ingredient will she need?

ISBN: 9780170368186

3 A cordial bottle states that one part of cordial is needed for every seven parts of water. What volumes of cordial and water are needed to make up 24 L of drink for the school social?

4 The human body contains 10 bacterial cells for every one human cell. If a body contains 110 trillion cells in total, calculate the number of bacterial cells and human cells.

5 The ratio of people with the normal number of ribs to those with one extra rib is 20:1. In a town of 13 272 people, how many would be expected to have an extra rib?

6 Butter and cheese are produced in New Zealand in a weight ratio of 5:3. If a total of 864 000 tonnes of butter and cheese was produced last year, how much of each was produced?

7 Uncle Scrooge revises his will, yet again! This time he leaves his $38 000 in the ratio of four parts to the cat, one part to each of the four older nieces, and two parts to his favourite niece, the youngest. How much does each get?

8 The town plan requires that the ratio of floor area to total land area in a housing development is less than 0.8:1. If a section is 535 m^2, calculate the maximum floor area that a house can have.

9 The town plan also requires that new subdivisions have a quarter of the land area set aside as open space. The rest is to be used in the ratio of four parts high-density housing to three parts medium-density housing to one part low-density housing. If the area of a new subdivision is 56 ha, calculate the areas used for open space and each type of housing.

ISBN: 9780170368186

Ratio calculations where one part is given

Example 1: The ratio of people to cars in New Zealand is about 8:5. If there are 4 568 000 people, about how many cars are there?

Step 1: Calculate the **value of each part** by dividing the value given (4 568 000) by the number of parts (8) it represents: 4 568 000 ÷ 8 = **571 000**

Step 2: Calculate the **value of the share** by multiplying the number of parts (5) by the value of each part: 5 x 571 000 = 2 855 000

∴ there are about 2 855 000 cars in New Zealand.

Example 2: An orchard is being planted with apple and pear trees in a ratio of 6:5. If 174 apple trees have been planted, how many trees will be planted in total?

Step 1: Calculate the **value of each part** by dividing the value given (174) by the number of parts (6) it represents: 174 ÷ 6 = **29**

Step 2: Calculate the **value of the total amount** by multiplying the number of parts (11) by the value of each part: 11 x 29 = 319

∴ there will be 319 trees in the orchard.

Solve the following ratio problems.

1 There are about seven sheep for every person in New Zealand. If the population is now 4 568 004, about how many sheep are there in New Zealand?

2 The ratio of growth in length of fingernails to toenails is 4:1. If my fingernails grew 5.6 mm in the last month, how much did my toenails grow?

3 The ratio of dried fruit to nuts in a recipe is 5:2. If Susie has 140 g of nuts, how much dried fruit will she need?

ISBN: 9780170368186

4 The ratio of boys to girls in a class is 4:3. If there are 16 boys in the class, how many students are there in total?

5 Caterpillars have 4000 muscles. The ratio of muscles in caterpillars to muscles in humans is 20:3. How many muscles do humans have?

6 In the athletic sports, the ratio of entries in running, jumping and throwing events was 8:3:2. If there were 296 entries into the running events, how many entries were there in total?

7 In the movie ***The Wizard of Oz***, Judy Garland was paid $500 per week. The ratio of her salary to that of Toto, the dog, was 25:3. How much was the dog paid?

8 Moths and butterflies are called lepidoptera. The ratio of moth species to butterfly species is 35:6. If there are 24 000 species of butterflies, how many species of lepidoptera are there?

9 When tossing a coin, the ratio of tails to heads is actually 500:495 (the picture of the head weighs more, so it ends up on the bottom more often). Ethan tossed a coin until he got 200 tails. Estimate the total number of times that he tossed his coin.

10 The scale on a map states 1:50 000. If a section of track on the map is 4.9 cm long, how long is the real section of track (in m)?

11 A model car is made in the ratio 1:64. If the real car is 4.0 m long, how long is the model car (in cm)?

  ISBN: 9780170368186

Mini-task

A pet shelter currently houses 168 stray animals.

- These are in the ratio of 15 dogs, 8 cats, 1 rabbit.
- On average, dogs need 0.25 kg of dried food per day, cats need 0.10 kg, and rabbits need 0.15 kg.
- A 10 kg bag of WOOF dog food costs $92 plus GST. However, the shelter gets a 15% discount off the GST-inclusive price.
- A 10 kg bag of BOW WOW dog food costs $103.50 including GST, but the supplier of the food will sell it to the shelter at the GST-exclusive price.
- At the same time last year, the pet shelter had 84 dogs and the ratio of animals was 14 dogs, 9 cats, 1 rabbit.

In the space below, work out the following.

- How many dogs, cats and rabbits the pet shelter currently has.
- The weight of food needed to feed all the animals for one week.
- The cost to the shelter of each type of dog food.
- Which dog food you would suggest they buy, and why.
- How many cats and rabbits the shelter housed at the same time the previous year.

ISBN: 9780170368186

Proportion

- Rates are (or can be) written using the word 'per'.
- Be careful that you use the same units in both the statements below.

Example 1: 3 kg of carrots cost \$4.95. Calculate the cost of 5 kg of carrots.

Step 1: Write the information you are given in a brief statement:

3 kg cost **\$4.95**

> Because you are trying to find a price, put the **price** on the right.

Step 2: Underneath, write a parallel statement about what you want to know:

3 kg costs **\$4.95**

copy ↓

5 kg costs **\$4.95** x $\frac{5}{3}$ = \$8.25

> Multiply by either $\frac{3}{5}$ or $\frac{5}{3}$. Use $\frac{5}{3}$ here because the \$4.95 needs to be made **bigger**.

Example 2: Sinta sometimes walks to school, which takes 27 minutes at a speed of 4 kph. Otherwise she skateboards to school at 10 kph. How long should skateboarding to school take her?

Step 1: Write the information you are given in a brief statement:

At 4 kph she takes **27 min**

> Because you are trying to find a time, put the **time** on the right.

Step 2: Underneath, write a parallel statement:

At 4 kph she takes **27 min**

copy ↓

At 10 kph she takes **27** x $\frac{4}{10}$ = 10.8 min

> Multiply by either $\frac{4}{10}$ or $\frac{10}{4}$. Use $\frac{4}{10}$ here because the 27 min needs to be made **smaller**.

 ISBN: 9780170368186

Answer the following questions.

1 9 kg of apples cost $24.75. Calculate the cost of 5 kg of apples.

2 3 kg of mince cost $32.85. Calculate the cost of 7 kg of mince.

3 When Lizzie drove from home to town at an average of 48 kph, it took her 21 minutes. Her mum drove from home to town at an average of 42 kph. How long did it take her?

4 When Lizzie drove from home to the beach at an average of 72 kph, it took her 28 minutes. Her mum did the same trip in 24 minutes. What was her mum's average speed?

5 An ant weighs 0.0003 kg and can lift 0.015 kg. Adam weighs 72 kg. If he could lift the equivalent of an ant, what weight could he lift?

6 The average person speaks for just 10 minutes per day. How long does the average person speak in one hour?

7 A snail can crawl 49 m in one hour. How far can it crawl in one minute?

8 In the USA, 2.5 cans of spam are consumed per second. How many cans of spam are consumed in the USA in one hour?

ISBN: 9780170368186

9 An average hamburger weighs about 100 g and costs \$3.50. An average car weighs 1100 kg. How much would 1100 kg (one car weight) of hamburgers cost?

10 The scale on a map states 1 cm : 500 m. If a track is 4.5 cm long on the map, how long is the real track?

11 A 1.5 mm tall flea can jump 200 mm. Whina is 1.6 m tall. If she could jump the equivalent of a flea, how high could she jump?

12 An impulse travels along a nerve at 274 kph (274 000 m per 3600 s). How long does an impulse take to travel the 1.5 m from Siale's toe to her head?

13 The average person sheds 18.13 kg of skin over their 70-year life. How much skin does the average person shed in a year?

14 It took five painters $17\frac{1}{2}$ days to paint a house. How long should it take a team of seven painters to do the same job?

15 A team of six parents makes 480 lamingtons in $2\frac{1}{2}$ hours. If a team of five parents makes 560 lamingtons, how long should the team take?

16 Another team of four parents took 3 hours to make 600 sandwiches. If a team of five parents worked for $2\frac{1}{2}$ hours, how many sandwiches should they be able to make?

ISBN: 9780170368186

Exchange rates

- The value of the New Zealand dollar changes with time.
- Its value also changes in relation to other currencies.
- Changes in its value are important when calculating the cost of overseas goods or travel.
- You can use the same method as in proportion calculations.

Exchange rates used in the examples and exercises:

$NZ1 = €0.65 (euro)
$NZ1 = £0.49 (pounds sterling)
$NZ1 = $US0.74 (US dollar)

Example 1: Michelle takes $3000 spending money with her to the United States. How many $US is that worth?

$$\$NZ1 = \$US0.74$$
$$\therefore \$NZ3000 = \$US0.74 \times \frac{3000}{1}$$
$$= \$US2220$$

Example 2: Ali returns to New Zealand with €8500 earned while working in the Netherlands. How many $NZ will he get when he banks it?

$$€0.65 = \$NZ1$$
$$\therefore €8500 = \$NZ1 \times \frac{8500}{0.65}$$
$$= \$NZ13\,076.92$$

Solve the following exchange rate problems.

1 Convert $NZ2500 to €.

2 Convert $NZ2500 to $US.

3 Convert $NZ2500 to £.

4 Convert €150 to $NZ.

5 Convert $US600 to $NZ.

6 Convert £15 to $NZ.

ISBN: 9780170368186

Mini-task

Anita is determined to get fit. She will join a gym. She investigates prices and finds that:

- George's Gym charges $260 for a three month subscription, or $850 for a year.
- Freddie's Fitness will give her the first month free, and then it charges $85 per month.
- Annie's Action charges $480 for each six month subscription, but they pay the GST for the first six months.

She has found the model of running shoes that she wants, and discovers that she can buy them for:

- $245 at her local sports shop
- $US170 on the internet. Currently $NZ1 = $US0.7014.
- £GB101.50 on the internet. Currently $NZ1 = £GB0.4520.

As part of her fitness regime she is going to run 1 km on the first day, and she will increase the distance she runs by 10% per day.

Complete the table and use the space below to answer the following questions.

- Which gym will be cheapest for her if she joins for six months?
- Which gym will be cheapest for her if she joins for one year?
- Where should she buy her running shoes? Explain.
- How far does she run on day 7, and what is the total distance she covers during the first week?

	Six months	**One year**
George's Gym		
Freddie's Fitness		
Annie's Action		

ISBN: 9780170368186

Standard form

- Standard form is a way of writing either very large or very small numbers without writing all the place holders (0s).
- Standard form is also known as scientific notation.

Examples: $5\,430\,000 = 5.43 \times 10^6$ $0.0000543 = 5.43 \times 10^{-5}$

Complete the following table.

Power of ten	Fraction	Whole number or decimal
10^3		
	$\frac{1}{10}$	
		1 000 000
10^{-2}		
		0.0001
	$\frac{1}{100\,000\,000}$	
10^0		

Standard form ⟶ ordinary numbers

Numbers bigger than 1

Example: Convert 7.12×10^5 to an ordinary number.

Method 1: $7.12 \times 10^5 = 7\,1\,2\,0\,0\,0. = 712\,000$

5 ⇒ shift the decimal point 5 places to the **right**.

Method 2: Use your calculator: enter 7.12 Exp 5 = ⇒ 712 000

Means '**x** 10 to the power'. (Do **not** enter the **x** sign.)

ISBN: 9780170368186

Numbers smaller than 1

Example: Convert 7.12×10^{-4} to an ordinary number.

Method 1: 7.12×10^{-4} = 0 . 0 0 0 7 . 1 2 = 0.000712

-4 ⇒ shift the decimal point **4** places to the **left**.

Method 2: Use your calculator: enter 7.12 [Exp] -4 = ⇒ 0.000712

Means '**x** 10 to the power'. (Do **not** enter the **x** sign.)

Convert these to ordinary numbers.

1 9.4×10^{2} = __________

2 1.42×10^{5} = __________

3 2.69×10^{3} = __________

4 6.111×10^{6} = __________

5 8.0×10^{0} = __________

6 4.5×10^{-2} = __________

7 9.467×10^{-1} = __________

8 3.7×10^{-5} = __________

9 6.2×10^{-3} = __________

10 1.83×10^{-8} = __________

Ordinary numbers ⟶ standard form

Numbers bigger than 1

Example: Convert 634 000 to standard form.

Step 1: Shift the decimal point to the **left** until there is exactly **one** non-zero digit on its left: 6 . 3 4 0 0 0 ⇒ 6.34

(This will give you a number between 1 and 10)

Step 2: Multiply by 10 to the power of however many places the decimal point was moved (**5**): 6.34×10^{5}

 ISBN: 9780170368186

Numbers smaller than 1

Example: Convert 0.00068 to standard form.

Step 1: Shift the decimal point to the **right** until there is exactly **one** non-zero digit on its left: 0 0 0 0 6 . 8 ⇒ 6.8

(This will give you a number between 1 and 10)

Step 2: Multiply by 10 to the **negative** power of however many places the decimal point was moved (-4): 6.8×10^{-4}

Check your answers by reversing the process.

Beware: To be in standard form, a number **must** be in the format **a.b x 10^{something}**

For instance: **7.4** in standard form is **7.4×10^0**
5 in standard form is **5.0×10^0**

Convert these to standard form.

1 543 =

2 1200 =

3 74.2 =

4 1.689 =

5 7 673 000 =

6 80 050 =

7 366 600 000 =

8 1 =

9 0.51 =

10 0.00067 =

11 0.014 =

12 0.000007 =

13 0.0001 =

14 0.00832 =

15 0.001101 =

16 0.0000000004 =

Calculations using standard form

Use your calculator to find the following answers. Write all answers in standard form.

1 $17(1.9 \times 10^5) =$

2 $(1.2 \times 10^7) \div 4.9 =$

3 $(3.8 \times 10^2) \times (2.995 \times 10^6) =$

4 $(1.3 \times 10^3)^2 \times (1.8 \times 10^7) =$

5 $(5.3 \times 10^{-4}) \div 6 =$

6 $(4.8 \times 10^8) \div (1.6 \times 10^2) =$

7 $7.4 \times 10^{-3} \div 500 =$

8 $(9.7 \times 10^8) + (1.0 \times 10^6) =$

9 $1000 - (1.45 \times 10^2) =$

10 $4.3 \times 10^{-5} \div 1.8 \times 10^{-2} =$

11 A ream of paper consists of 500 sheets, and is 52 mm high. Calculate the thickness of one sheet of paper.

12 About 5×10^8 tweets are sent per day. 77% of these are from outside the USA. How many tweets come from inside the USA per day?

13 The mass of a dust particle is about 7.53×10^{-7} g. About how many dust particles are there in the 50 g of dust that I emptied from my vacuum cleaner?

14 There are about 2.5×10^{22} molecules in one cubic centimetre of air. Each time you breathe, you take in about 5×10^2 cubic centimetres of air. About how many molecules do you take in with each breath?

15 Light travels one metre in about 3×10^{-9} seconds. How long would it take for light to travel the 493 km from Auckland to Wellington?

16 A handful of soil contains about 1.0×10^6 yeasts, 2×10^5 moulds and 1×10^4 protozoans. About how many yeasts, moulds and protozoans is that in total? (Note: As well as those there are about 1.0×10^{10} bacteria!)

ISBN: 9780170368186

17 Saturn is about 1.4×10^{12} m from earth. The distance that light can move in a year is known as a light year, and it is about 9.5×10^{15} m. Calculate how many light years Saturn is from earth.

____

18 One googol = 1×10^{100}. How many millions are there in one googol?

____

19 There are about 1×10^6 ants per person on earth. If the population of New Zealand is currently 4.575×10^6, estimate the number of ants in New Zealand.

____

20 1 mL of liquid can hold 5×10^8 bacteria. How many bacteria could there be in a 200 mL glass of liquid?

____

21 Your body produces 2.5×10^7 new cells per second. How many new cells could your body have produced in the last hour?

____

22 Your ribs move about 5×10^6 times per year. Estimate the number of times your ribs have moved in the last 24 hours.

____

23 In January 2015, the world population was 7.28×10^9 people. New Zealand's population was 4.5×10^6 people. Calculate the percentage of the world's population that lives in New Zealand.

____

24 The total width of an average plant cell is 5×10^{-2} mm. Its cell walls are 2×10^{-4} mm thick. Calculate the distance between the internal surfaces of the walls.

____

25 The mass of one bacterium is 9.5×10^{-13} g. The mass of one red blood cell is 2.7×10^{-11} g. How many bacteria would weigh the same as one red blood cell?

____

ISBN: 9780170368186

Practice tasks

Practice task one

The trip to Australia

Kaho, Kate and Krystal are raising funds for a school trip to Australia.

The trip has been planned for several years, so two years ago they had each put $500 into various saving schemes:

Kaho's paid him 4% compound interest annually.
Kate's paid her a total of 8% at the end of the two-year period.
Krystal's paid her 2% compound interest, paid at six-monthly intervals.

They all sell ice creams on sports day. They bought 150 ice creams, which cost them $2.80 each. They sold them all and made a 25% profit, which they shared equally between them.

Kaho made fudge. The ingredients cost him $23. He made 48 pieces, but he and his friends ate $\frac{3}{16}$. He sold the rest to the staff at morning tea time for $1.50 per piece. He kept the profit from the fudge.

Together they deliver telephone books. They receive $200 but Kaho works for seven hours, Kate for five and Krystal for four hours. They split the money in proportion to how long each worked.

Calculate how much each has raised towards the trip.

State clearly any assumptions that you make and explain your reasoning. You should show all your working clearly.

ISBN: 9780170368186

They each need a minimum of \$AU1500, and will need to change their New Zealand currency into Australian dollars. Over the last month the exchange rate has varied between \$NZ1 = \$AU0.96 and \$NZ1 = \$AU0.83. For each person, calculate how much more money (\$NZ) they still need to raise.

ISBN: 9780170368186

Practice task two

The netball tournament

Eleven friends are planning to enter a netball tournament. It will be held in a town that is 234 km from where they live. They will hire a van, stay away one night in a motel, and have dinner at a restaurant. Both the tournament venue and the restaurant are within walking distance of the motel. They will each need to buy a team T-shirt.

The van will cost them $245.00 plus 10% for insurance.

The van does 9.5 L per 100 km. Fuel costs 186.9c per litre.

The motel holds up to five people per unit. It costs $95 per night for two people, plus $15 per head for extra people.

Their meal will cost them $35 each, but if they book in advance the restaurant will give them a discount of 12%.

T-shirts will cost them $45 each including GST. However, because they are being supplied by a friend, he will pay the GST.

If the costs are shared between the 11 friends, how much will they each need to pay?

State clearly any assumptions that you make and explain your reasoning. You should show all your working clearly.

 ISBN: 9780170368186

After they have all paid their contributions, they discover that because they will be representing their school, the school will pay $\frac{2}{7}$ of the total cost. Calculate how much refund each of the friends should receive.

ISBN: 9780170368186

Practice task three

The barbecue

Sixty students are hosting a Cultural Awards barbecue for 240 people.
They need to buy sausages, bread, tomato sauce and cordial.

One kilogram of sausages will feed six people.
Sausages are $11 per kg, and the butcher will give the students a 15% discount.
One loaf of bread will feed 10 people. Bread costs $1.50 per loaf.
One bottle of tomato sauce will be enough for 15 people, and costs $4.50 per bottle.
The cordial costs $3.50 per 750 mL bottle. It needs to be diluted in the ratio
1 part cordial: 10 parts water. Each person needs 200 mL of cordial drink.

The school's Board of Trustees (BOT) will pay 20% of the cost of the food and drink.
The Parents Association will pay $\frac{1}{3}$ of the cost of the food and drink.
The 60 students will share the balance of the cost of the food and drink.

Calculate exactly how much each student will have to pay to just cover the cost of the food.

State clearly any assumptions that you make and explain your reasoning. You should show all your working clearly.

ISBN: 9780170368186

On the day of the barbecue, and after all the students have paid their $4, the organisers decide they should have fruit for afterwards. The fruit they select cost $59, but the owner of the shop said he would pay the GST portion of that price. The Parents Association couldn't afford to pay more, so the BOT said they would pay the extra cost. Calculate the percentage of the total cost of the barbecue that the BOT ends up paying.

ISBN: 9780170368186

Practice task four

Dave's kayaks

Dave is starting up a kayaking company. To do so he must purchase the necessary equipment, which includes kayaks, skirts, paddles and life jackets. He gets quotes from two different companies.

He requires:

12 kayaks
20 paddles
8 skirts
22 life jackets

Calculate the cost of each of these options and recommend the best value to Dave. He wants to buy all his equipment from one supplier.

Kayaks Galore

The local store can provide:

Kayaks for \$125 each
Paddles for \$12 each
Skirts for \$35 each
Life jackets for \$49 each

These prices all exclude GST but since Dave is a local, the store will give him 10% off the total price.

Cheap Kayaks Ltd

This larger chain can provide:

Kayaks for \$134 each
Paddles for \$15 each
Skirts for \$27 each
Life jackets for \$51 each

These prices all include GST but there is a '$\frac{2}{5}$-off discount' on the life jackets. However, there is a \$250 transport fee on each order.

From which shop will his equipment be cheaper?

State clearly any assumptions that you make and explain your reasoning. You should show all your working clearly.

 ISBN: 9780170368186

Cheap Kayaks have decided it will also give Dave one item free for every five similar items bought. For Kayaks Galore to match this, what discount would they need to offer?

Answers

All non-integral intermediate answers are rounded to a maximum of 4 sf.
All non-integral final answers are rounded to a maximum of 4 sf, or 2 dp for money.

Order of operations (pp. 7–8)

1	2	**2**	10
3	10	**4**	2
5	32	**6**	$\frac{9}{8}$
7	69	**8**	9
9	$\frac{7}{4}$	**10**	5
11	64	**12**	125
13	8	**14**	3
15	2	**16**	4
17	4	**18**	22
19	5	**20**	7
21	7	**22**	$3

Using your calculator (pp. 9–10)

Once upon a time ...

Once upon a time there was a **belle** named **Ellie**. **Eligible** princes would **besiege** her castle, which was located on a **hill** in the middle of a swamp. They would **slosh** and **slog** through **bogs** filled with **eels**, until the **soles** of their **shoes** were caked with **globs** of **hellish ooze**.
Ellie kept pet **geese**, which would **gobble** their food, then **loll** and **lie** about the palace. They would **hiss** at the princes.
However, the sight of Ellie would make the princes **boggle**, and dream of eternal **bliss**.
They would **beg** for her hand. But she would just **giggle** and reject them. They would **sigh** and **sob**, and turn on their **heels** and depart.
Prince **Billie** was different. He brought a bunch of **lillies** and an **ibis egg**. '**Hello**,' she said. '**Gosh**, none of the other **sillies** brought me an **ibis egg**. I can **see** that you're no **bozo**. You can stay and **boogie** with me.'
And they lived happily ever after.

Factors, multiples and primes (pp. 11–13)

Factors (p. 11)

1 1, 2, 4
2 1, 2, 3, 4, 6, 12
3 1, 3, 5, 15
4 1, 2, 3, 4, 6, 9, 12, 18, 36
5 1, 7
6 1
7 8: 1, 2, 4, 8
12: 1, 2, 3, 4, 6, 12
HCF: 4
8 5: 1, 5
20: 1, 2, 4, 5, 10, 20
HCF: 5
9 18: 1, 2, 3, 6, 9, 18
60: 1, 2, 3, 4, 5, 6, 10, 12, 15, 20, 30, 60
HCF: 6
10 11: 1, 11
21: 1, 3, 7, 21
HCF: 1
11 16: 1, 2, 4, 8, 16
30: 1, 2, 3, 5, 6, 10, 15, 30
HCF: 2
12 17: 1, 17
3: 1, 3
HCF: 1
13 He can make 12 bags, which each contains 2 mint lollies and 3 fruit lollies.
14 He can make 6 bags, which each contains 4 mint lollies, 6 fruit lollies and 5 caramel lollies.

Multiples (p. 12)

1 2, 4, 6, 8, 10, 12
2 5, 10, 15, 20, 25, 30
3 9, 18, 27, 36, 45, 54
4 12, 24, 36, 48, 60, 72
5 10, 20, 30, 40, 50, 60
6 1, 2, 3, 4, 5, 6
7 2: 2, 4, 6, 8, 10, 12
3: 3, 6, 9, 12, 15, 18
LCM: 6
8 5: 5, 10, 15, 20, 25, 30
3: 3, 6, 9, 12, 15, 18
LCM: 15
9 6: 6, 12, 18, 24, 30, 36
9: 9, 18, 27, 36, 45, 54
LCM: 18
10 8: 8, 16, 24, 32, 40, 48
12: 12, 24, 36, 48, 60, 72
LCM: 24

ISBN: 9780170368186

11 10: 10, 20, 30, 40, 50, 60
12: 12, 24, 36, 48, 60, 72
LCM: 60

12 7: 7, 14, 21, 28, 35, 42, 49, 56, 63
9: 9, 18, 27, 36, 45, 54, 63
LCM: 63

13 The 40th bead.

14 The 120th bead.

Primes (p. 13)

1	2, 3, 5, 7	**2**	11, 13, 17, 19
3	23, 29, 31, 37	**4**	41, 43, 47, 53, 59

Mixing it up (p. 13)

1	2, 3	**2**	23
3	3, 5	**4**	2, 3, 5
5	11 friends	**6**	On Monday 22nd
7	23 members with a subscription of $40		

Integers (pp. 14–15)

1	-12	**2**	2
3	0	**4**	10
5	4	**6**	7
7	-6	**8**	-3
9	9	**10**	5
11	-25	**12**	±5
13	-6	**14**	-1
15	-$13	**16**	13 °C
17	-8 °C	**18**	2
19	17°	**20**	6254 m
21	8 m	**22**	817 years
23a	30 BC	**b**	18 years old

Powers and roots (pp. 16–18)

Powers (p.16)

1	36	**2**	256
3	-8	**4**	81
5	-9	**6**	125
7	$\frac{1}{4}$	**8**	0.0625
9	5^2	**10**	3^3
11	-2^5 or $(-2)^5$	**12**	10^3
13	10^6	**14**	0.1^2

Fractional powers — roots (p. 17)

1	±4	**2**	±5
3	5	**4**	±2
5	±0.5	**6**	$\pm\frac{1}{2}$

Square root	Upper/lower limits	Value
$\sqrt{40}$	6 and 7	6.325
$\sqrt{6}$	2 and 3	2.449
$\sqrt{80}$	8 and 9	8.944
$\sqrt{67}$	8 and 9	8.185
$\sqrt{20}$	4 and 5	4.472

Negative powers (p. 18)

1	$\frac{1}{2}$	**2**	$\frac{1}{100} = 0.01$
3	$\frac{1}{16}$	**4**	$\frac{1}{1\,000\,000} = 0.000001$
5	$\frac{4}{5}$	**6**	3
7	$\frac{1}{4}$	**8**	3

Mixing it up (p. 18)

1	18	**2**	192
3	-16	**4**	-16
5	16	**6**	12
7	$-\frac{1}{25}$	**8**	4
9	9	**10**	145
11	48	**12**	72

Rounding (pp. 19–25)

1 Rounding to decimal places (pp. 19–21)

1	6.82	**2**	2.47
3	64.1	**4**	142.1
5	9.091	**6**	91.446
7	399.90	**8**	0.03
9	0.00	**10**	1.02
11	3.0	**12**	5.6
13	9.43	**14**	2028
15	8.3	**16**	2.851
17a	38 °C	**b**	38.2 °C
18a	5094 m^2	**b**	5090 m^2

Money (p. 21)

1	$57.86 ($57.90)	**2**	$7.13 ($7.10)
3	$9.09 ($9.10)	**4**	$12.43 ($12.40)
5	$61.97 ($62.00)	**6**	$8.97 ($9.00)
7	$5.13 ($5.10)	**8**	$31.53 ($31.50)
9	$72.95 ($73.00)		
10a	$45.80	**b**	$45.79

2 Rounding to significant figures (pp. 22–24)

Number	Number of significant figures
51.2	3
13 706	5
14.028	5
94 778.5	6
10	1
3.20046	6
4 461 030	6
12 000	2
1.20	3
1205	4
0.2	1

Continued over page

ISBN: 9780170368186

Number	Number of significant figures
0.004	1
43.50	4
9.670420	7
53.060	5
2.0670	5
79 000 001	8
10 000	1
60.0850	6
0.060410	5
1.030	4
0.090	2

1 53 **2** 35 000
3 0.068 **4** 10 000
5 110 **6** 0.00106
7 742 000 **8** 0.00694
9 1 672 000 **10** 0.07000
11 70 **12** 90 000
13 0.5 **14** 0.06
15 0.1 **16** 0.008
17 0.942 **18** 0.108
19 2.17 **20** 1.79
21 Lowest: 10.50 + 11.50 = 22.00
Highest: $11.4\dot{9} + 12.4\dot{9} = 23.\dot{9}$

Estimation (p. 25)

1 19 or 18.9 **2** 20
3 14 or 16.8 **4** 18
5 1 **6** 1 or 0.9
7 45 **8** 5 or 5.3
9 4 **10** 25
11 14 kg

Fractions (pp. 26–35)

1 $\frac{3}{4}$ **2** $\frac{1}{3}$
3 $\frac{1}{5}$ **4** $\frac{3}{5}$
5 $\frac{1}{3}\left(\frac{2}{6}\right)$ **6** $\frac{3}{8}$
7 $\frac{3}{4}$ **8** $\frac{1}{3}\left(\frac{2}{6}\right)$

1  **2** 
3  **4** 
5 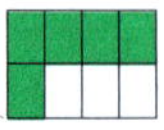 **6** 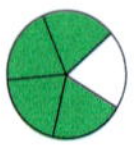
7  **8**

Converting between improper fractions and mixed fractions (p. 27)

1 $1\frac{1}{2}$ **1** $\frac{5}{2}$
2 $3\frac{1}{2}$ **2** $\frac{7}{4}$
3 $1\frac{1}{4}$ **3** $\frac{13}{4}$
4 6 **4** $\frac{10}{9}$
5 $2\frac{3}{5}$ **5** $\frac{14}{3}$
6 $3\frac{1}{4}$ **6** $\frac{11}{2}$
7 $3\frac{1}{3}$ **7** $\frac{69}{10}$
8 $1\frac{4}{8}$ or $1\frac{1}{2}$ **8** $\frac{11}{8}$
9 $6\frac{3}{4}$ **9** $\frac{23}{3}$

Equivalent fractions (p. 28)

1 $\frac{1}{3}$ **1** 32
2 $\frac{1}{3}$ **2** 9
3 $\frac{3}{4}$ **3** 105
4 $\frac{9}{10}$ **4** 5
5 $\frac{3}{5}$ **5** 5
6 $\frac{9}{10}$ **6** 39
7 $\frac{1}{4}$ **7** 12
8 $\frac{1}{2}$ **8** 11
9 $\frac{3}{4}$ **9** 2
10 $\frac{7}{12}$ **10** 120

Calculations involving fractions (pp. 29–30)

1a $1\frac{1}{4}$ **b** $1\frac{7}{15}$ **c** $\frac{21}{40}$ **d** $1\frac{1}{8}$
e $3\frac{7}{20}$ **f** $1\frac{22}{35}$ **g** $3\frac{1}{2}$ **h** 3
i $2\frac{1}{20}$ **j** $5\frac{1}{56}$

2a $\frac{1}{4}$ **b** $\frac{7}{15}$ **c** $\frac{1}{8}$ **d** $\frac{4}{15}$
e $1\frac{5}{6}$ **f** $\frac{13}{36}$ **g** $1\frac{5}{12}$ **h** $\frac{11}{20}$
i $2\frac{49}{60}$ **j** $1\frac{1}{380}$

3a $\frac{2}{15}$ **b** $\frac{1}{10}$ **c** $\frac{7}{20}$ **d** $\frac{1}{7}$
e $\frac{7}{15}$ **f** $\frac{5}{6}$ **g** $\frac{8}{9}$ **h** $3\frac{4}{15}$
i $19\frac{1}{4}$ **j** $\frac{1}{18}$

ISBN: 9780170368186

4a $\frac{3}{10}$ **b** $3\frac{1}{2}$ **c** $1\frac{1}{5}$ **d** 7

e 6 **f** $10\frac{2}{3}$ **g** $2\frac{7}{25}$ **h** $\frac{9}{20}$

i 30 **j** $\frac{1}{30}$

5a $\frac{1}{6}$ **b** $\frac{5}{24}$

5 $\frac{7}{12}$ **6** $\frac{7}{8}$ km

7 $\frac{1}{6}$ **8** $5\frac{5}{12}$ hours

9 $\frac{1}{4}$ kg **10** $\frac{5}{9}$

11 $\frac{4}{5}$ **12** $1\frac{11}{12}$ kg

Fractions and recurring decimals (p. 31)

1 0.666 ... **2** 0.474747 ...
3 0.5919191 ... **4** 0.416416 ...
5 5.11242424 ... **6** 10.13666 ...
7 10.13636363 ... **8** 10.13636636636 ...
9 $0.\dot{7}$ **10** $0.\dot{4}\dot{5}$
11 $9.\dot{1}\dot{4}$ **12** $2.3\dot{7}\dot{5}$
13 $20.0\dot{2}\dot{3}$ **14** $9.\dot{5}4\dot{6}$
15 $1.\dot{0}0\dot{1}$ **16** $0.\dot{2}63\dot{4}$
17 $0.\dot{6}$ **18** $0.\dot{4}\dot{5}$
19 $0.8\dot{3}$ **20** $0.1\dot{0}\dot{9}$
21 $0.\dot{2}9\dot{6}$ **22** $0.15\dot{9}\dot{0}$
23 $0.0\dot{1}6\dot{2}$ **24** $0.\dot{0}6\dot{0}$

Deciding which of two fractions is larger (p. 32)

1 **$0.\dot{3}$** 0.3 **2** 0.25 **$0.2\dot{6}$**
3 **0.5714** $0.\dot{5}\dot{4}$ **4** **$0.\dot{6}$** 0.6154
5 **0.4** 0.3846 **6** **$0.791\dot{6}$** 0.75
7 $0.8\dot{3}$ **0.85** **8** **$0.\dot{9}\dot{0}$** 0.9
9 $\frac{20}{30} = 0.\dot{6}$, $\frac{7}{11} = 0.6\dot{3}$, so better in science.
10 Supermarket 43c, so cheaper at market stall (40c).
11 Local \$9.99, so cheaper at city cinema (\$9.38).
12 5 kg \$2.20/kg, so cheaper in 1.5 kg bag (\$1.86).

Fractions of an amount (p. 33)

1 52 kg **2** 252 mL
3 \$44.50 **4** 38 km
5 24 m **6** 295 cm
7 102.3 g **8** \$248
9 70.8 kg **10** 34%
11 316 L **12** 58 km
13 568 cm **14** $\frac{3}{8}$
15 $0.08\dot{6}$ **16** \$10.58 each
17 \$73.50 for the family **18** $\frac{1}{4}$
19 $\frac{2}{15}$ **20** $\frac{3}{8}$

Applications of fractions (p. 34)

1 $\frac{2}{5}$ **2** $\frac{7}{15}$

3 $\frac{13}{24}$ **4** $2\frac{1}{12}$

Mini-task (p. 35)

Total cost:	Pizza	\$39.95
	Ice blocks	\$44.94
	Juice	\$47.76
	Mandarins	\$ 9.65
	Lollies	\$12.97
	Total:	**\$155.27**

∴ Costs \$6.4695 each, so round this to \$6.50

Temperature rise: 1 °C – -4 °C = + 5 °C

Ice block flavour: $1 - \frac{1}{4} - \frac{1}{3} = \frac{5}{12}$ were orange.

Highest common factor of 24, 30 and 42 is 6. So there will be 6 bags, each containing 4 butterscotches, 5 mints and 7 toffees.

Converting between decimals, fractions and percentages (p. 36)

Fraction	Decimal	Percentage
$\frac{1}{4}$	**0.25**	**25%**
$\frac{3}{4}$	0.75	**75%**
$\frac{3}{10}$	**0.3**	30%
$\frac{1}{5}$	0.2	**20%**
$\frac{3}{5}$	**0.6**	**60%**
$\frac{17}{25}$	**0.68**	68%
$\frac{3}{25}$	0.12	**12%**
$\frac{7}{8}$	**0.875**	**87.5%**
1	**1.0**	100%
$2\frac{1}{2}$	2.5	**250%**
$\frac{1}{25}$	**0.04**	4%
$1\frac{1}{5}$	1.2	**120%**

Percentages (pp. 37–41)

Calculating percentages (p. 37)

1 56% **2** $26.\dot{6}$%
3 45% **4** 88.10%
5 17.65% **6** $16.\dot{6}$%
7 E $19.\dot{4}$%, M 49.1%, A 25.9 %, N 5.6%
8 Private cars 57.89%, motor bikes 23.68%, taxis 10.53%, delivery vans 7.89%
9a 12.5%

ISBN: 9780170368186

b 2.44%

Percentages of an amount (p. 38)

1 53 km
2 124 mm
3 $130.50
4 276 mL
5 30.6 g
6 1.56 kg
7 25.62 cm
8 $15
9 $480
10 7.68 mL
11 $13.23
12 $4.13
13 10.08 km
14 $175
15 $5400
16 45 games
17 $15
18 78 students
19 nieces get $4560 each, cat gets $15 200

Increasing by a percentage (p. 39)

1 $54
2 127.5 km
3 788.8 cm
4 0.25 mm
5 921.6 m
6 70.08 g
7 18.41 g
8 607.5 L
9 116 km
10 3.3 min
11 0.051 kg
12 187 g
13 144 km
14 29.7 kg
15 80.16
16 $4719.60
17a $383.80
b $2494.70

Decreasing by a percentage (p. 40)

1 $73.50
2 238 m
3 97.5 mm
4 40.32 cm
5 1.44 L
6 10.8 kg
7 2.75 cm
8 177.6 L
9 510.3 g
10 44.16 min
11 1.274 tonnes
12 4.4 g
13 0
14 76.05 km
15 $45.77
16a $389.35
b $94.25
17 $334.38

Finding a percentage increase or decrease (p. 41)

1 Increase of 5%
2 Decrease of 10%
3 Increase of 15%
4 Decrease of 30%
5 Decrease of 25%
6 Increase of 2%
7 Decrease of 1%
8 Increase of 125%
9 Increase of 625%
10 Decrease of 86%
11 She lost 24%
12 It increased by 12.47%

GST (pp. 42–44)

Adding GST (p. 42)

1 $98.90
2 $17.83
3 $103.44
4 $117.67
5 $4665.54
6 $65.26
7 $0.26
8 $74.98
9 $609.50
10a $4025.00
b $525.00
11a $3.99
b $0.52

Calculating the pre-GST price and GST (p. 43)

1 $85.22
2 $23.87
3 $95.64
4 $223.29
5 $0.86
6 $852.74
7 $1.38
8 $75.39
9 $773.91
10 $652.04
11 $0.45
12 The salesman deducted 15% of the full price ($5999), which is $899.85, resulting in a sale price of $5099.15. He should have deducted 15% of the GST-inclusive price ($5216.52), which is $782.48, producing a sale price of $5216.52. So he sold the car too cheaply.

Mixing it up (p. 44)

1 6 students
2 $955.65
3 4.1%
4 $2587.50
5 13.5 km
6 $16.\dot{6}\%$
7 $123.15
8 $339.15
9 $5\frac{1}{2}$ cups
10 Science 83.3%, Maths 82%, History 82.61%. ∴ Science
11 $\frac{1}{3}$ pizza
12 14.93%
13 $2.60
14 20%
15 Cat $13 300, youngest $6460, other nieces $4560

Interest (pp. 45–47)

Simple interest (SI) (p. 45)

1 $525
2 $24 750
3 $1400
4 5.5%
5 6.25%
6 4 years

Compound interest (CI) (p. 46)

1 $8427.00
2 $30 198.74
3 $100 943.01
4 a $4416.96
b $396.28 more

Mini-task (p. 47)

Money withdrawn at the end of	Mum's scheme ($)	Dad's scheme ($)
one year	2100	2100
two years	2205	2200
three years	2315.25	2300
four years	2431.01	2400

He should choose his mother's scheme because apart from the first year when both give the same, his mother's scheme gives him more. After four years it gives him $31.01 more.

ISBN: 9780170368186

Watch costs	\$699 – 25% = \$524.25
Cricket bat costs	\$399 + 15% = \$458.85
Tablet costs	\$499 ÷ 1.15 = \$433.91
Sports trip costs	\$500
Total:	**\$1917.01**

Amount remaining = \$482.99

Ratios (pp. 48–53)

Simplifying ratios (pp. 48–49)

1	3:2	**2**	10:7:3
3	4:5	**4**	2:3
5	7:5	**6**	6:4:3
7	1:5	**8**	1:3:2
9	2:3:6	**10**	24:1
11	6:1	**12**	4:7:1

Ratio calculations where the total amount is given (pp. 49–50)

1 Sam 34 lollies : Sarah 17 lollies
2 1 cup liquid : $1\frac{1}{2}$ cups dry ingredients
3 3L cordial : 21 L water
4 100 trillion bacteria cells : 10 trillion human cells
5 632 people
6 540 000 tonnes butter : 324 000 tonnes cheese
7 Cat \$15 200, older nieces \$3800, youngest niece \$7600
8 428 m²
9 Open space: 14 ha
High-density housing: 21 ha
Medium-density housing: 15.75 ha
Low-density housing: 5.25 ha

Ratio calculations where one part is given (pp. 51–52)

1	31 976 028 sheep	**2**	1.4 mm
3	350 g dried fruit	**4**	28 students
5	600 muscles	**6**	481 entries
7	\$60 per week	**8**	140 000 species
9	398 tosses	**10**	2450 m
11	6.25 cm		

Mini-task (p. 53)

The animal shelter currently has 105 dogs: 56 cats and 7 rabbits.

Weight of food/day:

dogs:	26.25 kg
cats:	5.60 kg
rabbits:	1.05 kg
total:	32.90 kg ∴ in one week they need 230.3 kg

10 kg bag of WOOF GST-inclusive price =
\$92 x 1.15 = \$105.80

Less 15% discount = \$105.80 x 0.85 = \$89.93.

10 kg bag of BOW WOW costs
\$103.50 ÷ 1.15 = \$90

∴ WOOF dog food is very slightly cheaper, by \$0.07 per 10 kg bag. This difference is likely to be insignificant, and other factors such as whether the dogs like it, ease of supply, reliability, etc. could determine which food they buy.

At the same time the previous year, the shelter housed 54 cats and 6 rabbits, so 144 animals in total.

Proportion (pp. 54–56)

1	\$13.75	**2**	\$76.65
3	24 min	**4**	84 kph
5	3600 kg	**6**	$0.41\dot{6}$ min
7	$0.81\dot{6}$ m	**8**	9000 cans
9	\$38 500	**10**	2250 m (2.250 km)
11	$213.\dot{3}$ m	**12**	0.01971 s
13	0.259 kg	**14**	$12\frac{1}{2}$ days

15 $2\frac{1}{2} \times \frac{6}{5} \times \frac{560}{480} = 3\frac{1}{2}$ hours

16 $600 \times \frac{5}{4} \times \frac{2.5}{3} = 625$ sandwiches

Exchange rates (pp. 57–58)

1	€1625	**2**	\$US1850
3	£1225	**4**	\$NZ230.77
5	\$NZ810.81	**6**	\$NZ30.61

Mini-task (p. 58)

	Six months	One year
George's Gym	\$520	\$850
Freddie's Fitness	\$425	\$935
Annie's Action	\$417.39	\$897.39

Annie's Action is cheapest by \$7.61 for a six month subscription.
George's Gym is cheapest by \$47.39 for a one year subscription.

Shoes:

Local sports shop:	\$245
ex USA:	\$242.37
ex GB:	\$224.56

∴ They are cheapest by \$17.81 from Great Britain.

ISBN: 9780170368186

Distances run:

Day 1:	1 km	
Day 2:	1.1 km	
Day 3:	1.21 km	
Day 4:	1.331 km	
Day 5:	1.464 km	
Day 6:	1.611 km	
Day 7:	1.772 km	
Total:	9.488 km	

Standard form (pp. 59–63)

Power of ten	Fraction	Whole number or decimal
10^3		**1000**
10^{-1}	$\frac{1}{10}$	**0.1**
10^6		1 000 000
10^{-2}	**$\frac{1}{100}$**	**0.01**
10^{-4}	**$\frac{1}{10\,000}$**	0.0001
10^{-8}	$\frac{1}{100\,000\,000}$	**0.00000001**
10^0		**1**

Standard form ⟶ ordinary numbers (pp. 59–60)

1 940
2 142 000
3 2690
4 6 111 000
5 8.0
6 0.045
7 0.9467
8 0.000037
9 0.0062
10 0.0000000183

Ordinary numbers ⟶ standard form (pp. 60–61)

1 5.43×10^2
2 1.2×10^3
3 7.42×10^1
4 1.689×10^0
5 7.673×10^6
6 8.005×10^4
7 3.666×10^8
8 1.0×10^0
9 5.1×10^{-1}
10 6.7×10^{-4}
11 1.4×10^{-2}
12 7.0×10^{-6}
13 1.0×10^{-4}
14 8.32×10^{-3}
15 1.101×10^{-3}
16 4.0×10^{-10}

Calculations using standard form (pp. 62–63)

1 3.23×10^6
2 2.449×10^6
3 1.138×10^9
4 3.042×10^{13}
5 $8.8\dot{3} \times 10^{-5}$
6 3.0×10^6
7 1.48×10^{-5}
8 9.71×10^8
9 8.55×10^2
10 $2.3\dot{8} \times 10^{-3}$

11 1.04×10^{-1} mm thick
12 1.15×10^8 tweets
13 6.640×10^7 particles
14 1.25×10^{25} molecules
15 1.479×10^{-3} s
16 1.21×10^6 yeasts, moulds and protozoans
17 1.474×10^{-4} light years
18 1×10^{94}
19 4.575×10^{12} ants
20 1×10^{11} bacteria
21 9×10^{10} cells
22 1.370×10^4 times
23 6.181×10^{-2} %
24 4.96×10^{-2} mm
25 2.842×10^1 bacteria

Practice tasks (pp. 64–71)

Practice task one (pp. 64–65)

Kaho

Savings:	$540.80	($500 x 1.04^2)
Ice creams:	$35.00	($105 ÷ 3)
Fudge:	$35.50	(48 x $\frac{13}{16}$ x $1.50 – 23)
Phone books:	$87.50	($200 x $\frac{7}{16}$)
Total:	**$698.80**	

Kate

Savings:	$540.00	($500 x 1.08)
Ice creams:	$35.00	($105 ÷ 3)
Phone books:	$ 62.50	($200 x $\frac{5}{16}$)
Total:	**$637.50**	

Krystal

Savings:	$541.22	($500 x 1.02^4)
Ice creams:	$35.00	($105 ÷ 3)
Phone books:	$50.00	($200 x $\frac{5}{16}$)
Total:	**$626.22**	

$AU1500 @ $NZ1 = $AU0.96 = $NZ1562.50
$AU1500 @ $NZ1 = $AU0.83 = $NZ1807.23

Kaho needs
At worst: $NZ1807.23 – $NZ698.80 = $NZ1108.43
At best: $NZ1562.50 – $NZ698.80 = $NZ863.70

Kate needs
At worst: $NZ1807.23 – $NZ637.50 = $NZ1169.73
At best: $NZ1562.50 – $NZ637.50 = $NZ925.00

Krystal needs
At worst: $NZ1807.23 – $NZ626.22 = $NZ1181.01
At best: $NZ1562.50 – $NZ626.22 = $NZ936.28

ISBN: 9780170368186

Practice task two (pp. 66–67)

Van: $245 + 10% **= $269.50**

Fuel: Total km = 2 x 234 = 468 km
Fuel used = $\frac{468}{100}$ x 9.5 = 44.46 L
Fuel cost = 44.46 x $1.869 **= $83.10**

Motel: Need 3 units @ $95 each = $285.00
5 extra people @ $15 = $75.00
Total: **$360.00**

Meal: 11 x $35 – 12% **= $338.80**

T-shirts: Cost $\frac{11 \times \$45}{1.15}$ **= $430.43**

Total: = $1481.83
Each must pay: $134.71
(∴ $135)

With contribution from the school:

School pays: $1481.83 x $\frac{2}{7}$ = $423.38
Refund will be: $423.38 ÷ 11 = $38.49
(so $38.50)

Assumptions:

1. They did no extra km in the van.
2. The fuel consumption rate was accurate.

Practice task three (pp. 68–69)

Sausages: need $\frac{240}{6}$ kg = 40 kg @ $11 per kg
= $440
Less 15% = **$374**

Bread: need $\frac{240}{10}$ loaves = 24 loaves @ $1.50
= **$36**

Sauce: need $\frac{240}{15}$ bottles = 16 bottles
∴ need to buy 16 bottles @ $4.50
= **$72**

Cordial: need 0.2L x 240 = 48L
Dilute 1:10 → 750 mL makes 8.25 L
∴ need $\frac{48}{8.25}$ = 5.81 bottles
∴ need to buy 6 bottles @ $3.50
= **$21**

Total cost: $503.00
BOT pays $503 x 20% = $100.60
Parents pay $\frac{\$503}{13}$ = $167.67
Balance to be paid by students = $234.73
∴ $3.91 each, so charge $4 per head.

Including fruit:

Cost of fruit: $\frac{\$59}{1.15}$ = $51.30

Total cost: $503 + $51.30 = $554.30
BOT must pay: $100.60 + $51.30
= $151.90
% paid by BOT $\frac{\$151.90}{\$554.30}$ x 100 = **27.40%**

Assumptions:

1. No other costs to be covered, e.g. gas for barbecue.
2. Exactly 240 people come.
3. All 60 hosting the barbecue pay $4.

Practice task four (pp. 70–71)

Kayaks Galore

Kayaks: 12 @ $125 = $1500.00
+ GST = $1725.00
– 10% = $1552.50

Paddles: 20 @ $12 = $240.00
+ GST = $276.00
– 10% = $248.40

Skirts: 8 @ $35 = $280.00
+ GST = $322.00
– 10% = $289.80

Life jackets: 22 @ $49 = $1078.00
+ GST = $1239.70
– 10% = $1115.73

Total: $3206.43

Cheap Kayaks

Kayaks: 12 @ $134 = $1608.00

Paddles: 20 @ $15 = $300.00

Skirts: 8 @ $27 = $216.00

Life jackets: 22 @ $51 = $1122.00
$-\frac{2}{5}$ = $673.20

Transport: + $250.00

Total: $3047.20

∴ **Cheap Kayaks is cheaper by $159.23**

Cheap Kayaks' prices with one free item for every five bought:

Kayaks:	10 @ \$134 = \$1340.00
Paddles:	17 @ \$15 = \$255.00
Skirts:	7 @ \$27 = \$189.00
Life jackets:	19 @ \$51 = \$969.00 $-\frac{2}{5}$ = \$581.40
Transport	+ \$250.00

Total: \$2615.40
Kayaks Galore would have to give a 19% discount so that \$3206.43 – 19% = \$2597.21, which is \$18.19 cheaper than Cheap Kayaks.

Assumptions:

1. Goods were of equivalent quality at both shops, and could be supplied within similar periods.
2. With the one-free-for-five deal at Cheap Kayaks, the two fifths off life jacket price still applied.

ISBN: 9780170368186